THE EQUICENTRAL SYSTEM SERIES BOOK

G000161593

Horse Property Planning and Development

Jane Myers and Stuart Myers
Equiculture Publishing

Copyright © September 2015

ISBN: 978-0-9941561-9-8

Email: stuart@equiculture.com.au

Disclaimer

The authors and publisher shall have neither liability nor responsibility to any person or entity with respect to any loss or damage caused or alleged to be caused directly or indirectly by the information contained in this book. While the book is as accurate as the authors can make it, there may be errors, omissions and inaccuracies.

About this book

It does not matter if you are buying an established horse property, starting with a blank canvas or modifying a property you already own; a little forward planning can ensure that your dream becomes your property. Design plays a very important role in all our lives. Good design leads to better living and working spaces and it is therefore very important that we look at our property as a whole with a view to creating a design that will work for our chosen lifestyle, our chosen horse pursuit, keep our horses healthy and happy, enhance the environment and to be pleasing to the eye, all at the same time.

Building horse facilities is an expensive operation. Therefore, planning what you are going to have built, or build yourself is an important first step. Time spent in the planning stage will help to save time and money later on.

The correct positioning of fences, laneways, buildings, yards and other horse facilities is essential for the successful operation and management of a horse property and can have great benefits for the environment. If it is well planned, the property will be a safer, more productive, more enjoyable place to work and spend time with horses. At the same time, it will be labour saving and cost effective due to improved efficiency, as well as more aesthetically pleasing, therefore it will be a more valuable piece of real estate. If the property is also a commercial enterprise, then a well-planned property will be a boon to your business. This book will help you make decisions about what you need, and where you need it; it could save you thousands.

Thank you for buying this book and please consider either leaving a review or contacting us with feedback, stuart@equiculture.com.au.

About the authors

Jane Myers MSc (Equine Science) is the author of several professional books about horses including the best selling book **Managing Horses on Small Properties** (published by CSIRO).

Jane has lived and breathed horses from a young age and considers herself to be very fortunate in that she has been able to spend her life riding, training and studying these amazing animals.

Stuart Myers (BSc) has a background in human behaviour and has been a horse husband for more years than he cares to remember.

Jane and Stuart are particularly interested in sustainable horsekeeping practices and issues, such as low stress horse management that also delivers environmental benefits. They present workshops to horse owners in Australia, the USA and the UK about sustainable horse and horse property management as part of their business, **Equiculture**.

Their experience is second to none when it comes to this subject as they keep up with recent advances/research and are involved in research themselves. They travel the world as part of their work and they bring this information to you via their books, online resources and website.

See the **Equiculture website** www.equiculture.com.au where you will find lots of great information about sustainable horsekeeping and please join the mailing list while you are there!

Jane and Stuart also have another website that supports their **Horse Rider's Mechanic** series of workbooks. This website is www.horseridersmechanic.com why not have a look?

Photo credits

All photos and diagrams by Jane Myers and Stuart Myers unless otherwise accredited. Any errors and omissions please let us know.

Contents

Introduction																																	1

## Chapter 1: Horse housing/holding facilities												3
### Surfaced holding yards																		4
Holding yard size and shape																7
Holding yard surface																						10
Holding yard fences																						14
Innovative additions to a surfaced holding yard				16
### Shade and shelter																						16
Your climate																										17
Shade and shelter building materials											24
Shade and shelter surface																	24
Shade and shelter size																				24
Innovative additions to a shade/shelter										25
### Stables																											26
Why stables?																									26
Stable designs																								31
Stable size																										37
Stable roofs																										39
Stable walls and partitions																41
Stable windows																								47
Stable doors																										49
Stable floors																									51
Stable fittings																								54
### Other facilities																						57

## Chapter 2: Fences and gates																65
### Fence and gateway safety																68
### Fence visibility																						70
### Fence dimensions																						70
### Fence types																								73
Hedges																												75
Timber fences																								76
Pipe and steel fences																				79
Stone fences/walls																						81
Mesh fences																									82

PVC/vinyl fences 85

Wire fences 86

Electric fences 88

Fence posts **97**

Tread-in posts 97

Fibreglass posts/rods 98

Insultimber posts 98

Other hardwood posts 98

Softwood timber posts 98

Composite/recycled/plastic posts 99

Steel posts ('star pickets') 99

Droppers/battens/stays **100**

Gates and gateways **103**

Gates and gateway safety in particular 103

Gateway dimensions 108

Gate types 109

Chapter 3: Riding arenas and training yards **111**

Do you really need one? **114**

Can this area be multi-purpose? **115**

Riding arena, training yard or both? **117**

Indoor or outdoor? **117**

All-weather surface size and shape **119**

Base and surface **120**

Fencing your all-weather surface **129**

All-weather surface fence height 130

All-weather surface fence materials 131

All weather surface gateways 135

All-weather surface lights **137**

All-weather surface maintenance **137**

Chapter 4: Horse facility planning **141**

Making a plan **141**

Building permits 144

Options for construction 145

The planning framework **151**

The environmental factors 151

The horse welfare factors 154

Your budget 157
The ergonomic factors 159
The safety and security factors 160
The natural elements 163
The aesthetic factors 167
Planning horse property infrastructure **169**
The house and garden 169
Property access 173
    Laneways 175
Horse facility positioning 177
    Shade and shelter positioning 178
    Fence and gateway positioning 182
    All-weather surface positioning 189
Manure management planning 193
Water management planning 197
    Sources of water 197
    Storage of water 200
    Using water 203
    Planning for clean water 208
Vegetation planning 210
    Windbreaks and firebreaks 210
    Revegetation of steeper land 212
    Easy areas to increase vegetation 212

Appendix: The Equicentral System **219**
How The Equicentral System works **219**
    Additional information 220
    The Equicentral System in practice 223
The Equicentral System benefits **225**
Horse health/welfare benefits: 225
Time saving benefits: 227
Cost saving benefits: 228
Safety benefits: 230
Land/environmental management benefits: 230
Public perception benefits: 232
Manure and parasitic worm management benefits: 233
Implementing The Equicentral System **235**
On your own land 235

On small areas of land   235
On large areas of land   237
In different climates   238
Using existing facilities   239
On land that you lease   241
On a livery yard (boarding/agistment facility)   242
With single horses in 'private paddocks'   243
Starting from scratch   244
Minimising laneways   245
Temporary laneways   247
Constructing a holding area   249
Constructing a shade/shelter   249
Fencing considerations   249
Management solutions   **250**
Feeding confined horses   250
Changing a horse to 'ad-lib' feeding   253
Ideas for extra exercise   258
Introducing horses to herd living   259
The Equicentral System - in conclusion   **261**

Further reading - A list of our books   **262**
Recommended websites and books   **267**
Bibliography of scientific papers   **267**
Final thoughts   **267**

Introduction

Building horse facilities is an expensive operation. Therefore, planning what you are going to have built, or build yourself, is an important first step. Time spent in the planning stage will help to save time and money later on. This book will help you to make the correct decisions and spend money on the right things. It is a guide, but make sure you do lots of other research as well; you can never have too much knowledge.

This book is about a *practical* approach to planning and building horse facilities. Remember - your 'dream' may be your horses' 'nightmare'. It is important that you learn as much as possible about horse behaviour before committing to any expensive projects. When planning and building any horse facilities, it is also good idea to talk to other people who have carried out similar projects and see what you can learn from them.

This book starts with extensive information about **horse housing/holding facilities** e.g. surfaced holding yards, shade/shelters and then stables. The next chapter covers **fences and gates/gateways**. After that comes **riding arenas and training yards**. Then it is on to the **planning** section of the book. The book includes information about **The Equicentral System**, a total equine management system that allows you to manage your horse/s, your land, the environment *and* your lifestyle in a sustainable way - a win-win situation all round.

Good planning leads to a beautiful horse property that is enjoyable for all.

Chapter 1: Horse housing/holding facilities

If you plan to practice good land management, then you need areas for holding horses. Horses naturally spend around 12–16 hours a day grazing. That means that if horses are on the land 24 hours a day, they are also standing around/playing etc. ('loafing') and sleeping for 8–12 hours a day. Without horse holding areas, these other behaviours are taking place on your precious pasture, which results in compacted soil and its associated problems – less biodiversity, less pasture/more weeds/soil erosion/mud/dust etc. Therefore, problems are being created for the future which will take time and money to fix.

If you want to practice good land management, then you need areas for holding horses.

Horse holding areas generally take the form of surfaced holding yards with shade/shelters and/or stables, with or without yards attached. In addition to reducing unnecessary hoof pressure and its resultant wear and tear on the land, these areas are also useful for giving individual horses supplementary feed and for tacking horses up for work etc.

The sorts of questions that you need to answer before you build or make changes to your existing horse housing/holding facilities are:

- What do you *really* need?
- What would *your horses* prefer?
- What are the *priorities*?
- What can you *afford*?

- Can you *amalgamate* facilities for better use of funds/space?
- What will be *cost effective*?
- What will make the task of caring for horses *easier*?

Surfaced holding yards

Surfaced holding yards are called various names in different parts of the world. In the US these areas are commonly called 'dry lots'. In the UK they are sometimes called 'turn out areas', although they are still quite rare despite being desperately needed in such a wet climate. In Australia and New Zealand, like the UK, they have been slow to catch on to date. At Equiculture, we like and use the term 'surfaced holding yards'.

Traditional horse properties have stables for horses, but these are not usually the best type of facility for keeping horses.

—

See *The Equicentral System Series Book 1 – Horse Ownership Responsible Sustainable Ethical* for a discussion about how and why stables evolved. See also the section *Why stables?* in this book.

—

Surfaced holding yards are vital in order to manage the grazing pressure that horses can inflict during dry or wet times of the year.

Surfaced holding yards are vital in order to manage the grazing pressure that horses can inflict during dry or wet times of the year. They also allow you to manage the pasture intake of horses at the times of the year when pasture is abundant, therefore saving some for later when it may not be growing as well or not at all, and when the horses need supplementary feed.

Surfaced holding yards allow you to manage these booms and slumps so that your horses do not end up obese or underweight and your land does not end up degraded. This strategy will help you to grow healthy 'happy' pasture.

Surfaced holding yards can be used to *vastly* reduce the amount of pressure your horses inflict on the land, without compromising their behaviour or necessarily reducing their time spent grazing. However, if you do need to reduce their time spent grazing, surfaced holding yards allow you to do this.

There are so many benefits to keeping horses off the land when they are not actually grazing. These include:

- Better mud management, in fact, mud should become a thing of the past.

- Better dust management. Dust is related to mud, no mud – no dust.

- Fewer or even no instances of skin conditions caused by mud e.g. goodbye to greasy heel/mud fever etc.

- Better weed seed management by confining weed seeds from hay, something that is especially important if that hay is bought in rather than made on the property.

- Better manure management. Manure is easier to pick up from surfaced areas and this collected manure can be composted and put back on the land to increase the organic matter of your soil.

- More pasture over time – by holding your horses at the right time, the pasture gets in front of the horses rather than lagging behind the horses. You will therefore have healthier, 'happier' plants.

- Increased biodiversity - by removing horses before they overgraze the plants, the less robust plants get chance to set seed and thrive.

- Increased opportunity for cross-grazing and all the benefits that this provides - e.g. better weed management, better parasitic worm management, a reduction of pasture plant wastage due to the 'roughs' and 'lawns' that are created when horses only graze a pasture. This also means that you can produce your own meat if you wish.

—

See *The Equicentral System Series Book 2 - Healthy Land, Healthy Pasture, Healthy Horses* for information about land/pasture and manure management.

—

Better mud management, in fact, mud should become a thing of the past.

Individual holding yards or one larger surfaced holding yard can be built for pairs or even groups of horses. If you have more than one herd to accommodate, this larger surfaced holding yard can be replicated for each herd.

A large surfaced holding yard can be integrated into a system of management whereby the paddocks are linked back to a central area to which horses can take themselves, see Appendix: ***The Equicentral System*** for information about our total land management system. Since these areas cost money to set up, it is worth considering how you can maximise their use; with a bit of extra thought and planning, a large surfaced holding yard can also be used for riding or for stock work – see the section ***Can this area be multipurpose?***

A horse property usually benefits from having surfaced holding yards, even if there are stables already on the property. Stables are often too hot to use during summer (depending on the materials used to build them and the climate) and it is during the heat of the day that horses most need shade. At this time of day, horses like to stand around swishing their tail at flies and dozing, behaviour which causes compaction to your land over time. Surfaced holding yards can either be attached to the stable complex if you are building one, or be separate to them.

If the property has stables and the stables are used regularly, it makes sense to have outside yards attached directly to them if possible, as opposed to having them situated in a separate area. This will save a lot of time moving horses between the two and save doubling up on buildings e.g. building stables *and* shelters, because the surfaced holding yards will need a shelter. However, keep in mind that on a hot day, an enclosed stable may be too hot to retreat to (from an attached yard), so some outside shade may still be required. This could be something as simple as 'shade sails' attached to the roof of the building.

A horse property usually benefits from having surfaced holding yards, even if there are stables already on the property.

As mentioned before, stables are, in most cases, unnecessary and the same money can instead be put to better use building surfaced holding yards that with good shelter. This type of surfaced holding yard works well on its own, without the need for stables; stables alone, on the other hand are not ideal. These surfaced holding yards are then comfortable for horses to use at *any* time of the day throughout the year and vastly reduce the amount of wear and tear on the land.

Surfaced holding yards are invaluable in terms of land management and plenty of thought should go into their construction.

Holding yard size and shape

A common mistake is to make *individual holding* yards that are either too large or too small. If they are too large they are expensive to build and maintain e.g. more surface material and fencing is required and they use up too much space that could otherwise be dedicated to growing pasture. Keep in mind that unless horses are grazing and/or with other horses, they do not tend to move much at all, therefore providing them with a very large area (that is not under pasture) does not necessarily mean that they will get essential exercise. Horses do not understand the concept of 'keeping fit' and will only move if there is a reason (from their point of view) to move.

If the property has stables and the stables are used regularly, it makes sense to have outside yards attached directly to them if possible.

—

See **The Equicentral System Series Book 2 - Healthy Land, Healthy Pasture, Healthy Horses.**

—

If surfaced holding yards are too small or narrow, the horse will be *too* confined and unable to roll safely or move around comfortably. In addition, if the horse is positioned next to another yarded horse, one horse can intimidate the other, even if there is a fence between them. This is because supplementary feed *creates* this behaviour in domestic horses, not because horses are aggressive just for the sake of it.

A good size to aim for with *individual holding* yards is between 50sq.m to 100sq.m per holding yard (60sq.yds. to 120sq.yds.). A *surfaced holding yard* can be roughly multiples of 50sq.m (60sq.yds.) per horse. The corners are best rounded off to avoid horses becoming trapped by other horses.

Keep in mind that feeding concentrate supplements to horses causes them to be more competitive than when they are simply grazing or eating hay. In the confines of a small area, even the best of friends can injure each other at feed times (see Appendix: **Feeding confined horses**). If space and budget allows, a larger surfaced holding yard with individual holding yards attached directly to the outside gives the best of both worlds.

Think about what will be the optimum size for surfaced holding yards. A good size to aim for with individual holding yards is between 50sq.m to 100sq.m per holding yard (60sq.yds. to 120sq.yds.)

If you are also planning to use a riding arena or training yard as a surfaced holding yard, you will need to integrate the required size of both and go with whichever is larger. The corners can be easily rounded off by fixing a rail across each one.

If you are building (or already have) stables, keep in mind that long narrow yards, whilst not ideal, are sometimes unavoidable when they are attached to standard size stables. A possible option for surfaced holding yards that are attached to stables is to have one larger yard per two stables that two horses can use for part of each day/night on a rotational basis. If space and budget allows, have stables that are double the standard width, which then allows surfaced holding yards that are approximately 7m (24ft) wide. Yet another option if you are planning to build just four to six stables is to have them arranged in a U shape, giving more room around the outside for surfaced holding yards. Of course, if you are building surfaced holding yards *instead* of stables, you are then free to make them any size and shape you like.

A possible option for surfaced holding yards that are attached to stables is to have one larger yard per two stables that two horses can use for part of each day/night on a rotational basis.

Holding yard surface

The subsurface of surfaced holding yards can be permeable, but should preferably be impermeable, as this prevents the leaching of nutrients into any underground water (the water table). Compacted limestone will achieve this. In sensitive areas, for example near a watercourse, it may be necessary to lay a concrete base or use a rubber matting/concrete or limestone combination to prevent any leaching. If rubber mats are used, you will need to place absorbent bedding in one area of the surfaced holding yard to encourage urination and soak up urine; horses do not like to urinate on a hard surface as it then splashes on their legs. Other possibilities for a sub-surface are to compact the existing sub soil, after removing the topsoil, or excavate this sub soil and replace with layers of progressively finer material starting with rocks and ending up with fine gravel, compacting each layer as it is laid.

This sub surface will usually require a top surface that is more horse friendly than the subsurface might be, and also needs to be able to cope with wet weather.

Top surface materials can be various types of sand, sawdust, shavings, pine bark, shell grit, or fine gravel etc. Various types of rubber are also available that may be worth considering.

Top surface materials can be various types of sand, sawdust, shavings, pine bark, shell grit, or fine gravel etc.

When horses eat directly off sand, they inadvertently pick up grains of it along with the feed. The sand then collects in the gut and *can* lead to sand colic over time. Rubber mats create a better feeding area than sand so in this case, it can work well to partially surface a sand surfaced holding yard with rubber mats.

Yet another alternative is to use rubber for the whole surface of the yard. Either solid rubber 'pavers' or porous rubber 'pavers' can be used. These have various names as there are various manufacturers that produce them, you will need to do some homework and find out if they are available in your locality and if they will be cost effective and suitable for your climate etc.

—

Materials have different names from region to region, so ask around about what is available in your area. What you use will depend on factors such as your budget, the climate, the local availability of materials and the amount of shelter over the surfaced holding yard.

—

This is a porous rubber product used for the whole of the yard surface. In this case grass has been allowed to grow through it.

If you are building a riding arena or training yard that will also be used as a surfaced holding yard (as mentioned in the previous section), then whatever surface you decide on for that will also be fine for using as a surfaced holding yard, so let that take precedence (see the sections about surfaces in the relevant sections about riding arenas and training yards).

The surfaced holding yards should have a slight slope (2% to 4%) for surface runoff, but otherwise be fairly level, because horses should never be forced to stand on a significant slope for long periods of time. Horses can suffer from premature joint problems if kept in such conditions, because they have no sideways flexion in their leg joints; an evolutionary strategy to allow them to gallop over 'rough' surfaces. Remember - in the wild horses can move themselves if they are uncomfortable, whereas a domestic horse is unable to do this, so forcing them to stand on a slope is not good.

Runoff water from other areas should not be allowed to pass through the surfaced holding yards, so water collection tanks should be fitted to any shelters or stables and drainage channels should be installed around the surfaced holding yards to channel water around them, rather than through them.

A buffer zone of plants and bushes can be grown around the surfaced holding yard area and this vegetation will act as a filter for any runoff from the yards, whilst at the same time helping to retain the surface. 'Sleepers' or logs can also be used to help retain the surface, although vegetation can work just as well and is cheaper. Check that any plants are not poisonous; it is better if the horses cannot actually reach them even if they are safe to eat, because the horses will overgraze them and the plants will not be able to thrive.

Horses should never be forced to stand on a significant slope for long periods of time because they have no sideways flexion in their leg joints.

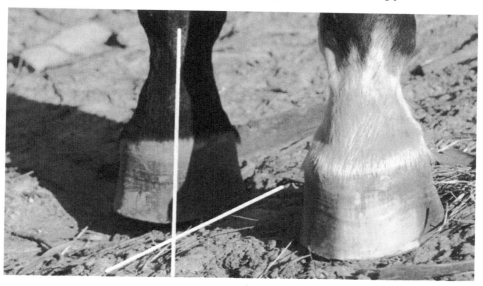

A buffer zone of plants and bushes can be grown around the surfaced holding yard area and this vegetation will act as a filter for any runoff from the yards.

Holding yard fences

All fences on a horse property should be **strong and safe** however, the smaller the area a horse is confined to, the more a horse comes into contact with the fence, and therefore this rule is especially true for fencing around surfaced holding yards.

Also the smaller the area horses are to be held in is, the **higher and stronger** the fence should be.

So, whereas a paddock fence can usually be 1.2m (3.9ft), a surfaced holding yard fence should usually be higher. A better height for surfaced holding yard fences is 1.4m (4.5ft) or higher. However, this is all very dependent on what you plan to do with your horses. Older, quieter horses are not as likely to challenge a fence as younger or insecure horses. If you are planning to hold a horse in a surfaced holding yard while, you take their companion for exercise for example, the surfaced holding yard may need a strong, high fence. Also, whereas a simple plain wire fence may be fine for a paddock, a more solid type of fence is usually required for a surfaced holding yard.

Whereas a simple plain wire fence may be fine for a paddock, a more solid type of fence is usually required for a surfaced holding yards.

See the section *Fence types* for more detailed information about each type of fencing mentioned below. In this section, the fence types are discussed with relevance to their suitability as a surfaced holding yard fence only.

Plain board fencing (post and rail) is a common type of fencing for surfaced holding yards, however the wood can splinter and cause injuries. Some types of

14

wood splinter more easily than others, so do some research about the types of wood you have available to you before deciding. Treated pine posts and rails are usually cheaper than hardwood, but may be poisonous if horses chew them. Wooden fencing also tends to be high maintenance and can soon start to look unkempt unless regularly serviced. Electric fencing is not a good idea for surfaced holding yards as it can result in the horse/s not being able to move freely without fear of touching the fence.

Steel fences, made from recycled steel pipe or commercial livestock steel fence panels are a very suitable option for surfaced holding yard fences and are *reasonably* safe when two horses are positioned on either side of a fence in individual holding yards. Horses should be able to interact reasonably safely with one another over the fence, providing each horse has enough space to avoid unwanted attention.

Mesh fencing should only be used in surfaced holding yards if it is strong enough for horses and the gaps are small enough to prevent a hoof from getting caught. Generally speaking, a commercial horse mesh fence is fine in this situation, but other types of mesh – such as 'dog fence' or 'ringlock' are not.

Surfaced holding yard gates should swing both ways, should lie flat against fence when open, be free of any projections and be wide enough to get machinery into the surfaced holding yard for topping up the surface and any maintenance work. See the section **Gates and gateways**.

The subject of horses interacting over a fence is tricky; if horses start to play over a fence, they can get into serious trouble. Keeping horses together will prevent this scenario and horses usually cause less damage to each other than fences do to horses (ask any horse vet!).

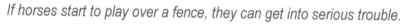

If horses start to play over a fence, they can get into serious trouble.

Innovative additions to a surfaced holding yard

Scratching posts of various sizes and shapes can be positioned so that horses rub on them. This measure reduces wear and tear on fences.

There are various commercial types of scratching posts available and some that can be home-made.

Shade and shelter

Horses need some form of shade and shelter to protect them from hot sun, strong winds and rain. In the domestic situation, horses usually have no choice about where they are at any given time (see Appendix: *The Equicentral System* for information about how you can redress this issue), therefore it is very important that a shelter is provided for them so that they can access it when necessary.

Most horses will use a shelter in order to escape flying insects and the sun in hot weather, rather than for protection from plain cold weather. In fact, cold weather alone does not usually bother horses if they are healthy; it is hot sun or driving wind and rain that they strive to escape from.

At the same time, millions of years of evolution have taught horses that to have no retreat is to be in danger, so for this reason a shelter that looks perfectly acceptable to the human eye due to being enclosed, warm and dry, may be virtually ignored by a horse. Aim to build a 'horse friendly' shelter that your horse/s will appreciate and use.

A shelter for horses does not have to be expensive.

Your climate

The building type should vary with the climate. Unfortunately, shelters are often built that do not suit the climate and are positioned incorrectly (see the section *Horse facility positioning*). In both scenarios, the horses will not use them unless they are forced, resulting in stressed horses and wasted money. Also keep in mind that, in some climates, shade is more important and in others, shelter from the elements is more important.

Horses are kept in various climates around the world, ranging from very hot to very cold and from very dry to very wet. Some climates have four distinct seasons (summer/spring/autumn/winter), whereas some only have two distinct seasons (wet season/dry season). Others have short summers and long winters with huge snowfalls. It is important that you understand your local climate and plan to provide the correct type of shelter for that climate. Some examples are given below:

Temperate climates - sun, cold wind, rain, sleet and snow:

- A man-made shelter, such as a large solid roof with minimal sides, combined with trees/bushes positioned to reduce the wind speed. In this case, the solid roof provides excellent shade in summer *and* protection from heavy rain. The vegetation acts as a windbreak, allowing cool breezes to pass through in summer and slowing strong winds in winter. While the vegetation is still maturing, you can use 'shade cloth' on the sides of the building as a temporary windbreak.

- A wall with no roof; a simple windbreak - this type of shelter suits temperate climates as long as the horses are fit and healthy. They would also need shade in summer, which could be provided by trees.

- In colder windy climates, at least one reasonably solid side is usually essential because wind travels sideways. However, the more open-sided the shelter, the more acceptable it is to most horses.

A wall with no roof; a simple windbreak.

In colder windy climates, at least one reasonably solid side is usually essential.

Tropical climates - hot, humid, dual season, wet/dry weather:

- The main considerations are protection from flying insects in summer, shade in summer and shelter from high (usually warm/hot) winds and very heavy rain in the wet season. High humidity is also a factor in these climates and horses struggle to cope with this.

- A simple roof without solid sides gives protection against sun and heavy rainfall without being too enclosed. Shelters that have two or three sides can be too hot. Anything that stops a breeze is detrimental, so the more 'open air' the shelter is, the better. The roof needs to be large enough (or have enough space around it) so that the horses can utilise it for sun protection when the sun is at all angles.

This roofed shelter provides four small paddocks with shelter and shade and allows the occupants to be able to see each other and stand near each other.

Arid climates - very hot and dry seasons followed by cold and dry seasons:

- The main considerations are shade and protection from flying insects in summer and shelter from hot drying winds and cold nights (sometimes very hot through the day and very cold at night).

- A 'shade sail' attached to a solid roof and/or poles which will provide shade and to some extent shelter. It needs to be constructed so that it can cope with high winds.

Arctic climates - short summers/very cold winters, deep snow for long periods:

- The main considerations are protection from flying insects in summer, shade in summer and very deep snow in winter.

19

- This climate warrants solid sided buildings more than other climates. A three sided building with an open front works well. The opening must be positioned so that it is not facing the wind.
- The roof should be constructed so that heavy piles of snow do not suddenly slide off the roof into the yard.

Some examples of shelters and their pros and cons:

- A portable structure such as a more traditional wooden shelter, but fitted with 'slides' or wheels so that can be dragged from paddock to paddock by tractor if necessary. These can become difficult to move as they age – not to mention that you need a tractor to move them. Rather than moving the shelter around the paddocks, think about getting the horses to walk back to the one shelter. See Appendix: *The Equicentral System*.
- A steel 'shade shed' that simply rests on the ground and can be dismantled reasonably easily and relocated. These shelters are not necessarily meant to be moved around the paddocks, but they are designed so that they can be dismantled and relocated to another property if you move. They can usually be erected without planning permission, as they are not usually regarded as a permanent building (check with your local authority). You may even be able to enclose some of the sides. A shade shed's heavy weight is largely what keeps it in place, but it can also be 'pinned' to the ground with steel pins if necessary, such as in areas with very high winds. This type of structure is common in the USA and Australia where their original use is as a 'car port', but they are now used for a variety of applications including livestock shelters.
- A fabric over steel frame structure – such structures are commonly in use in many countries. Again, they have various applications ranging from large grain storage/indoor riding arena, down to individual livestock shelters and hay storage etc. If used as a shelter for horses, they need to be open at either end so that they do not get trapped by other horses at the back. As with steel shade sheds, they may be able to be erected without planning permission.

A steel 'shade shed' that simply rests on the ground and can be dismantled reasonably easily and relocated.

A fabric over steel frame structure – such structures are commonly in use in many countries.

- Traditional 'paddock shelters' tend to have just one entrance/exit at the front which can mean that horses get trapped by other horses inside the shelter. For this reason, horses may be unwilling to stand inside it and will simply stand outside, using the walls as a windbreak – therefore this kind of shelter is often a waste of money. A further problem with this type of shelter is that it is difficult for sun and air to get in, so they tend to become muddy inside unless they have a proper surface.

Traditional 'paddock shelters' tend to have just one entrance/exit at the front which can mean that horses get trapped by other horses inside the shelter.

This is a good alternative to the previous style of shelter as it has an opening at the back.

Natural shelter

Horses actually prefer natural shelter such as trees and bushes, because they are not a fully enclosed. There are pluses and minuses with using vegetation for shelter for horses.

Trees and bushes need protection from horses; horses can be hard on natural vegetation both by chewing on it, and also by compacting the soil around the roots as they stand underneath the plant/tree. In order to survive, vegetation needs to be planted on the outside of any surfaced holding yards, away from direct pressure from horses. If horses are to stand over the area that the roots of a tree grow in, then thick mulch should be placed under the tree, as opposed to allowing horses to stand on bare soil over the tree roots.

Evergreen trees will provide shade and shelter all year round, whereas deciduous trees will provide shade in summer and, when they lose leaves in winter, they allow winter sun to pass through. However, they also allow winter wind and rain to pass through, so you need to think carefully about what you need from trees/bushes and select accordingly. When possible, select plants that are native to the area, as these will tend to be the easiest to grow (because they are in their natural environment) and will also provide habitat for native wildlife.

Aim to increase natural shelter in paddocks by planting trees and bushes around the outside of paddocks, because natural shelter has many other benefits besides simply providing shelter.

—

See the section *Some of the benefits of trees and bushes on a horse property* in *The Equicentral System Series Book 2 - Healthy Land, Healthy Pasture, Healthy Horses*.

—

Plant trees and bushes around the outside of holding yards.

23

Shade and shelter building materials

The considerations for shelter building materials are the same considerations as for stable building materials, and shelters, like stables, should be lined if they are made from materials that a horse can kick through. One difference is that the horse/s will have access to the *outside* of the building as well as the inside, so make sure that there are no external projections that can injure a horse. It must be strong as it *will* be rubbed by horses. The internal height of a shelter should be at least 2.75m (9ft), but very large horses would need it to be even higher.

Shelters, like stables, should be lined if they are made from materials that a horse can kick through.

Shade and shelter surface

A shelter that is positioned on bare earth will become muddy in wet conditions and dusty in dry conditions. In addition, if the sun cannot get to the area, it will stay muddy even when the surrounding area has dried out. So, a shelter usually needs a surface and this can be the same as that used for the enclosing surfaced holding yard if that is where it is positioned. Bedding can be added to this surface to encourage the horses to lie down.

Shade and shelter size

The shelter needs to be large enough inside so that the horses will be comfortable, especially if it has enclosed sides. See the section **Stable size** for information about standard sizes for individual stables and use this as a guide. If more than one horse is to use the shelter, it needs to be large enough so that they can all access it safely at the same time, and get in and out without being trapped.

Innovative additions to a shade/shelter

There are various things that can be added to a shelter to improve it such as:

- Shade sails, which can be added to a solid structure in order to relatively inexpensively increase the protective area.

- Fans, which can be placed inside shelters to discourage flying/biting insects.

- Plastic strips or strips of shade cloth, which can be hung from the front of a shelter. These will help to keep flying/biting insects out. The combination of these strips, together with fans, makes a significant difference to the impact that flying/biting insects have and reduces or eliminates the need for rugs, repellents etc.

Shade cloth hanging from the front of a shelter helps to keep flying/biting insects out.

Stables

The use of stables for horses is now a very contentious issue. Some people believe that horses should be free ranging, as nature intended, whereas others believe that stables are an absolute necessity when keeping domestic horses, particularly in an urban or business environment.

Why stables?

Stables have been in use for thousands of years; the earliest so far discovered were in ancient Egypt approximately 3000 years ago and housed the pharaoh's chariot horses. This early use of stables is typical of why they were developed; in historical times horses were the primary vehicle and were either ridden, or used as the 'engine' for the vehicles of yesteryear. They were vital in cities, for warfare and for agriculture. Everywhere in fact that nowadays motorised vehicles are used.

These horses spent many hours a day working and, at the end of each day, they needed an enclosure to rest and recuperate, eat concentrated feed and be ready to begin work again next day. At the same time the enclosures for these horses were small by necessity, because space is always at a premium in a city.

Another factor that led to the use of stables is that there are not enough hours in a day for a hard working horse to replenish its energy reserves by grazing. These horses were usually working for at least 10 hours a day, sometimes much more, so even if pasture was available, which it generally wasn't in a city, a horse needed supplementary nutrition in the form of concentrate feed.

These working horses needed to be easily accessible and the use of stables allowed working horses to be available at a moment's notice for whatever purposes their owners required. Welfare wasn't necessarily a concern, but convenience was.

In the military, stables also kept horses nearby so that they were available for work and warfare. In addition, elaborate stable routines evolved, particularly within military circles, as a means to keep young soldiers occupied when not at war.

Stables also evolved as a warm and sheltered environment for humans to work while taking care of horses. Stables allow people to handle and care for horses in relative comfort by protecting those people from the elements. Enclosed barn style buildings are a good example of this, with the central isle between two enclosed rows of stables and large doors at either end of the building. These stables particularly developed in parts of the world where the winters are extremely cold (Northern Europe).

So, we can see that stables were not really developed for the welfare of horses, but more for the convenience and comfort of their owners and handlers. Whilst there may still be a need for stables in modern times, we should be aware of the reasons *why* they developed and we should understand that these reasons may no longer be applicable in the 21st century (see also the section ***The horse welfare factors***).

We can see that stables were not really developed for the welfare of horses, but more for the convenience and comfort of their owners and handlers.

The common arguments *for* having stables:

- They provide a place to keep a horse clean, warm, dry etc.

- They can be used to house horses in a relatively small space. Therefore horses can be kept in areas – such as cities – where they could not otherwise be kept.

- Some horses actually 'like' being stabled e.g. some horse owners feel that because their horse stands at the gate waiting to be brought in then they must 'want' to be stabled.

- It is the 'traditional' way to keep horses.

The reasons outlined above are a good example of why we should look to the study of horse behaviour before deciding what we think our horse wants and needs.

—

See ***The Equicentral System Series Book 1 – Horse Ownership Responsible Sustainable Ethical*** for much more information about horse behaviour.

—

These reasons in favour of stabling can all be mitigated by other more rational reasons. For example:

- Yes, stables keep a horse clean, warm and dry, but so does a surfaced holding yard with a shelter (surfaced so that the horse is not standing in mud).

- Yes, they allow larger numbers of horses to be kept in smaller areas, but should we doing this anyway? Is it fair for an animal that evolved to be almost constantly on the move while grazing to be kept in a tiny space where movement is severely restricted? Again, if horses are to be kept in small areas, then a more 'horse friendly' solution is to provide an area that is larger than a stable, with a surface and a shelter, or better still, a surfaced holding yard with a larger shelter, so that horses can be inside *or* outside if they wish.

- Lastly, horses stand at the gate waiting to be let back in when they are fed concentrates. This gives the impression that they 'love' their stable when in fact it is the feed that they 'love'. They would just as happily stand in a surfaced holding yard with a shelter if their paddock gate was left open, allowing them to get back to it (see Appendix: *The Equicentral System*).

The common arguments *against* having stables:

- The physiology of the horse shows us that horses are healthier if they are outside as much as possible. Stables confine horses to a small space with poor air quality and challenge a respiratory system that is not designed for indoor living. Horses do need shelter from the wind and the rain, but horses actually thrive in cold climates (more so than hot climates).

- With good management, horses can be kept in a *relatively* small area and still be able to carry out many of their natural behaviours. Stables are usually the most expensive option when building horse facilities (rather than surfaced holding yards with shade/shelter) and may require more on-going maintenance depending on their construction, due to more building materials being used.

- Animal welfare protocols continue to develop in the western world, legislation is beginning to be passed that prevents or limits the use of stables. So, building stables might prove to be a costly mistake for the future. At the time of writing at least one European country is legislating against the use of stables.

- We need to remember that modern horses rarely, if ever, work as hard as their domesticated ancestors. Modern horses that are stabled full-time may spend only an hour a day 'in work', then spend most of the rest of their day in the stable and therefore, welfare issues arise. When you consider the relative size of a horse to the size of its stable, then compare that with enclosures for zoo animals, you can see where the concerns lie. Can you imagine the outcry if a zoo kept an animal as large as a horse in an enclosure as small as a stable?

28

This is what we as horse owners need to be aware of, as these horses do not spend adequate time moving.

Stables confine horses to a small space with poor air quality and challenge a respiratory system that is not designed for indoor living.

Over-confinement can lead to physical and behavioural problems in horses due to the highly unnatural environment that confinement provides. Confinement automatically reduces a horse's opportunities for movement and interaction with other horses. In addition, it often leads to a reduction of fibre in the diet. This does not *have* to occur, but there is a tendency for it to occur in a stabled horse. **Owners that stable their horse/s *tend* to feed a diet that is too low in fibre because:**

- They may be concerned about how much they are spending on hay, because when a horse is stabled, every mouthful of feed has to be paid for. This can lead an owner to try to economise by feeding less hay.

- Concentrated feed is easier to store, especially when space is a premium.

- Hay may be harder to source than concentrate feed, particularly in an urban environment.

- They may be concerned about their horse getting a 'hay belly'. It is a myth that a horse that eats a high fibre diet will develop a 'hay belly', but just like most myths this way of thinking prevails.

- They may simply not understand how important fibre is to a horse.

- There is strong marketing pressure to feed concentrate feed but no such marking pressure for feeding hay.

Horses do not develop a 'hay belly' just because they have a high fibre diet.

The physical problems that can occur when horses are housed and fed incorrectly include gastro-intestinal problems such as colic and gastric ulcers, which come about from a lack of adequate fibre. The behavioural problems include stereotypic behaviours such as 'wind-sucking' (also from a lack of adequate fibre) and weaving (from a lack of movement).

Confined horses should *at least* have daily turnout into a grassy paddock or, failing that, into a large surfaced area. In both situations, this should be with other horses so that they can all take part in social, behaviours such as mutual grooming sessions and play. They should also be given large amounts (preferably 'ad-lib') of **low-energy** (as opposed to small amounts of high energy) fibrous feed e.g. hay. Good planning and **good horse management** should ensure that stables *or* surfaced holding yards do not become dungeons for horses. By using them appropriately, stables have some good uses but then so do surfaced holding yards with shade/shelters, so make sure you think about the pros and cons of each option before deciding what to build if you are fortunate enough to be making this decision.

—

If a horse cannot live at pasture full time for whatever reason (not enough/too much grass, land too wet/too dry), then the next best thing for both the mental and physical health of a horse is a surfaced holding yard with a shade/shelter.

—

If you have decided that you need stables, you have a lot of planning to do. The first things to consider are how many stables are needed and what will they be used for? For example, will the horses be mainly outdoors or mainly indoors? Then decide what other areas you would also like to have. Then you can research what style of stables you would like and work from there.

If you have decided that you need stables, you have a lot of planning to do.

Stable designs

Check out as many designs as possible before building or deciding what to have built. The design and arrangement of facilities in the building can have a big influence on stable management; facilities which are well-planned and welldes-igned make horse care much easier and more enjoyable for all.

Allow space for movement of horses, wheel barrows etc. and maybe small items of machinery such as a mini tractor. As well as individual stables for horses, the building may include other areas such as a wash/vet area, tack storage, kitch-enette, laundry, feed storage, separate tie-up areas, workshop, toilet/s etc. (See the section *Other facilities*). **Some of the different styles of stables are:**

- A straight single row (row form) of individual stables with either no overhang to the roof, a roof overhang (for some protection from the elements) or a full veranda that can be used for various purposes such as for tacking up as well as protection from the elements. This style of stables can be positioned for optimal wind/rain and sun protection as all of the stables face the same direction.

A straight single row (row form) of individual stables.

- A double (back to back) row of stables with or without the same overhang/veranda options. Like with the single row variety, the horses have access to fresh air. Fewer building materials are used as the stables are back to back, but the downside of this style is that while one row may be facing the best way in terms of sun and wind/rain, the other might not.
- Other variations are L shaped or U shaped rows which increase ease of use and comfort for humans, without compromising on ventilation to the extent that an enclosed building (barn style) does.

L shaped stable blocks can be a good compromise.

- Barn style complexes where two rows of stables face each other with a central aisle (breezeway) down the middle. Barn style complexes (outside of America these are often referred to as 'American Barns' even though they originated in the colder parts of Europe) can have the door closed at one or both ends in winter to stop high winds, and open in summer to allow a breeze to pass through. Barn style complexes are very popular because they allow horses to be managed out of the weather; people can take care of horses in extremely cold winters in reasonable comfort. Keep in mind that the climate in northern Europe and America can be very different to other parts of the world such as Australia.

Barn style complexes are more costly to erect than individual stables that are in a row due to using more building materials. From the horses' point of view, they are probably the least desirable as they have the poorest air quality, because the individual stables are indoors rather than outdoors, and the poorest view, because the view is only of the inside of the building itself.

Barn style complexes originated in very cold climates such as Sweden.

These factors can be greatly improved by having a permanently open window or, better still, a stable door to the outside of each individual stable. If there is a door, a 'surfaced holding yard' (sometimes called a 'paddock' in certain parts of the US) can be attached to the back of each individual stable which means that each horse can at least get outside into the fresh air. A door on the back of each individual stable also improves the fire safety factor of a barn complex. In hot climates, these attached surfaced holding yards will also need shade, because depending on what it is constructed from, the inside the stable building will be too hot during the day. This shade could be in the form of a simple 'shade sail'.

Keep in mind that that just as with back to back rows, one side of the barn may end up poorly positioned in terms of sun and wind/rain.

There should be more than one exit to any enclosed building, and these exits should be as far apart as possible, at each end of the 'breezeway'. They should open easily for increased ventilation and fire safety. In longer, barn style buildings, there should also be exits at intervals down each side. The breezeway needs to be wide enough for machinery (for feeding/mucking out) and for moving horses safely. A good width is about 3.6m (12ft).

If there is a door, a 'surfaced holding yard' can be attached to the back of each individual stable which means that each horse can at least get outside into the fresh air.

—

See the **Equiculture website Equine Safety and Industry section** for links to Fire safety in stables - **www.equiculture.com.au/equine-safety-and industry.html**

—

In addition to deciding what type of stable building to build, you should also think about how you can build more 'horse friendly' stables. For example, 'traditional' books about building stables will usually advise you to build stables with full height partitions (from floor to ceiling).

Individual stables that have solid partitions from floor to ceiling have many disadvantages:

- They do not allow any interaction between occupants.
- They have reduced air flow.
- They are more expensive to build.

Therefore, solid walled stables are the least desirable and cannot be justified if horses spend significant amounts of time in them (e.g. if the horses do not spend a large amount of time outside each day, in the company of other horses). Stables built in this way are actually no better than cages in a zoo, which used to be a common sight before the zoo industry made significant changes to the way that animals in their care were kept. Even stallions (or you might say *especially* stallions) should be able to see and even touch other horses.

In hot climates, these attached surfaced holding yards will also need shade, because depending on what it is constructed from, the inside the stable building will be too hot during the day. This shade could be in the form of a simple 'shade sail'.

Source - Alayne Blickle of Horses for Clean Water - USA.

Stables should be open above the height of the 'kicking boards' (traditionally 1.2m/4ft), or meshed/barred above the 'kicking boards'. This latter arrangement at least allows some interaction between neighbouring horses and better airflow between each individual stable. Bars should be no more than 7cm/2.7 ins apart to reduce the risk of a rearing horse putting a hoof through them, which would probably lead to a catastrophic injury. Alternatively, partitions above solid 'kicking

35

boards' can be partially enclosed and partially open, so that there is an area where each horse can get behind at feed times and an area where neighbouring horses can interact with each other at other times.

If a stable has solid walls to above head height, consider removing part of the wall (above chest height) so that two horses have an area where they can interact. This does not have to be all the way along the dividing wall, just part way is fine if you wish. If you are building stables from scratch aim to build 'horse friendly' stables from the start.

'Horse friendly' stables.

—

See **The Equicentral System Series Book 1 – Horse Ownership Responsible Sustainable Ethical** for more information about horse behaviour and how you can make sure that you understand what is really important to a horse and how you can provide it to your horse.

—

Stable size

The traditional industry standard size for an individual stable is generally 3.65m x 3.65m (12ft x 12ft). Keep in mind that the industry standard for stables was developed a very long time ago in a time when horses worked exceptionally hard for a living and animal welfare was not necessarily considered an issue (see the section **Why stables?**). Remember - horses today rarely work anything like as hard as their ancestors and therefore a fully stabled horse will then spend 23 or more hours a day in an area that is incredibly small.

The traditional industry standard size for an individual stable is generally 3.65m x 3.65m (12ft x 12ft).

A stable for a pony should be *at least* 3m x 3m (about 10ft square). Large horses (e.g. more than 16.2hh) need larger stables e.g. 4.3m x 4.3m (about 14ft square) or the stable should open out on to a surfaced holding yard.

The amount of time that a horse is going to spend in the stable should determine the size of it. The figures described above are the **minimum** sizes for stables where horses are expected to spend more time than it takes to eat supplementary feed or are being held just before being ridden etc. Horses should not be expected to spend all night *and* all day in such a small space (see the section **The horse welfare factors**). If you *are* building stables, aim to make the individual stables as large as possible. Better still, aim to build surfaced holding yards with shelters (see the section **Surfaced holding yards**).

Smaller stables are fine if they are only for occasional or short term use. For example, if horses live outside but are brought in each day for supplementary feed and/or work. In this case, it is a good idea to have partitions that swing back against the wall, or that can be removed entirely, so that two small stables can become one large one if necessary. Then, if a horse *has* to spend more time inside, such as in the case of an injury requiring temporary confinement, the stables can be expanded.

It is a good idea to have partitions that swing back against the wall, or that can be removed entirely, so that two small stables can become one large one if necessary.

Stable roofs

The choice of roofing material includes any material that can be used for any domestic building. Factors to consider are aesthetics (including in relation to the other buildings), cost, strength, insulation properties, safety (including fire safety) and ease of maintenance.

There are many different types of roof material depending on the building style. Some of the options are tiles, shingles and corrugated steel. Corrugated steel is a common roof material and can be either silver coloured or have a bonded colour coating. Keep in mind that some local authorities do not allow silver coloured tin to be used for buildings as it can produce a glare in the sun.

Stables need at least 1 metre (3.3ft) of free air flow above the head level of each horse; therefore 2.75m (9 foot) is the recommended minimal internal roof height and higher is even better. Higher roofs give better ventilation and are cooler in the heat. The pitch of the roof is important with regards ventilation; a higher pitched roof that has ventilation outlets positioned high up will draw hot air upwards and outwards.

Higher roofs give better ventilation and are cooler in the heat.

Stables that are part of a stable complex can have poor ventilation in the middle stables compared to those on the ends. Open windows/top doors make a huge difference to air quality in this style of stables. In addition, the roof will need to allow air to flow out so that the air is changed regularly. The emphasis should be on fresh air flow rather than draughts.

Roofs can be insulated so that temperature extremes and condensation are reduced. This will also help to deaden noise in hailstorms and heavy rain.

Transparent panels provide extra light into buildings that are enclosed and reduce power for lighting usage; however they also let in heat. In hotter climates and even temperate climates that have hot sunny days in summer, use them only in the breezeway (rather than directly over stables) and limit their use to approximately one every 7m (23ft) of the length of the breezeway. Also, position them on the side of the building that gets more shade than sun to further reduce heat. If possible, transparent panels may be able to be placed in a position that let in winter sun but not overhead summer sun. Look at modern sustainable house designs for ideas.

With stable complexes, the roof can be a low or high loft style. High loft styles can have an upstairs floor which can be used for extra storage or even accommodation; check with the local authority with regards to regulations.

High loft styles can have an upstairs floor which can be used for extra storage or even accommodation.

Any roofs should be fitted with gutters and spouts, and the water diverted with surface or subsurface drainage so that it is taken away from the building without causing wet areas near the buildings footings and does not run across any surfaced holding yards. Water should be collected in water storage tanks whenever possible. The overflow from tanks will also need to have correct drainage to take surplus water away. See the section **Water storage tanks**.

Stable walls and partitions

The exterior of the building may determine the interior to some extent, for example some building materials will make up both the inside as well as the outside aspect, whereas some building materials require lining. Whatever material is used for the interior aspect should be strong, smooth and preferably should not splinter or dent if kicked.

Some of the materials that can be used in the construction of stable buildings are:

Corrugated steel

This is a popular building material that is usually available in a galvanised metal finish or coloured (in which case you can match it to other buildings on the property). It can be used for exterior walls and roofs.

Advantages of this material:

- Quick to erect.

- Neat and tidy finish.

- Relatively inexpensive.

- Relatively fire resistant.

Disadvantages of this material:

- It must be lined on the inside of the stable or shelter, otherwise a horse can put a hoof through it, which will result in injury.

- It does not have good insulation properties, therefore the building will be hot (inside) in hot weather and cold (inside) in cold weather. Insulation can be added to the building, but this will increase the cost of using corrugated steel so make sure you factor that in.

Flat steel sheets

This material can be used for individual stable partitions and as a liner for exterior walls that are made of wood or corrugated steel for example. The sheets must be strong enough to withstand the kick of a horse.

Advantages of this material:

- Neat and tidy finish.

- Low maintenance.

- Easier to keep clean than wood.

- Does not rot like wood.

- Relatively fire resistant.
- Chew resistant.

Disadvantages of this material:

- Not as attractive as more traditional materials such as wood.
- More expensive than wood.

This material can be used for individual stable partitions and as a liner for exterior walls that are made of wood or corrugated steel for example.

Timber

In general wood is not recommended in parts of the world that have a high risk of wood eating termites (white ants), although some timbers are termite resistant so do some research if you plan to use wood. If you are not sure, check with your local authority to find out what the termite risk is for your area. If a building is constructed of wood (exterior walls), it has good insulation properties and is therefore cooler in hot weather in particular. The material used for the roof will affect this factor greatly though.

Timber planks

Timber would have to be one of the most common and versatile building materials and of course one of the most traditional. **It can be used for:**

- The exterior walls of the building and will not require lining if It is thick enough (5cm/2 inch).
- The lining of a building when the exterior walls are made of materials such as corrugated steel.

- Partitions between individual stables etc.

If a building is constructed of wood (exterior walls), it has good insulation properties and is therefore cooler in hot weather in particular.

When used as a liner or for partitions, planks should be 5cm/2 inch thick to a height of at least 1.4m (4.6ft). The planks can either be fitted vertically or horizontally. They should be fitted flush to the floor so that a horse cannot get a hoof caught underneath them, however this means that the bottom plank, or the bottom edge of vertically fitted planks, will eventually rot unless protected in some way.

Advantages of this material:

- The most traditional looking material.
- Requires no particularly special skills or equipment to work with compared to certain other materials.

Disadvantages of this material:

- Can rot.
- Can be expensive depending on the grade.
- Is not fire resistant.

- Is not chew resistant, although it can be improved with the addition of steel covers in the areas that are likely to be chewed. However, if horses are chewing you need to check that their fibre requirements are being met.

Timber would have to be one of the most common and versatile building materials and of course one of the most traditional.

Timber sheets

These usually take the form of plywood sheets (of various thicknesses). There are many variations ranging from plain plywood to products that are treated so that they do not rot (such as 'Marine Ply'). The price obviously varies along with the standard. They can be used for the same applications as timber planks.

Advantages of this material:

- Neat and tidy finish.
- Fast and easy to put up.
- Relatively inexpensive.

Disadvantages of this material:

- If a sheet is damaged (e.g. kicked) the whole sheet will need to be replaced, rather than just one plank.

- Can rot.

- Is not fire resistant, although some types would be better than others.

- Is not chew resistant, although it can be improved with the addition of steel covers in the areas that are likely to be chewed. However, if horses are chewing you need to check that their fibre requirements are being met.

Timber sheets are fast and easy to put up.

Concrete

Concrete is a versatile material for building exterior walls or lining stables etc and can be painted or left unpainted. Concrete can be poured into moulds to make up the walls, eliminating the need for an inner lining, or pre-made concrete wall panels can be used. Another alternative is concrete blocks. Concrete can withstand 'rough' treatment, but will collect dust unless they are rendered. If a building is constructed of concrete (exterior walls), it has good insulation properties and is therefore cooler in hot weather in particular. The material used for the roof will affect this factor greatly though.

Advantages of this material:

- Neat and tidy finish.
- Strong.
- Low maintenance.

- Easier to keep clean than wood.
- Does not rot like wood.
- Relatively fire resistant.
- Chew resistant.
- Good insulation properties therefore cooler in hot weather in particular.

Disadvantages of this material:

- Not as attractive as more traditional materials such as wood, although when painted, as in the case of the poured concrete partitions in the previous picture '**Horse friendly stables**' (see the section **Stable designs**), this material is very attractive.

Other building products

Commercially produced, man-made 'weather boards' are long-lasting and more durable than timber. They are low maintenance, but more expensive than timber. These can be used as an outer wall only, as they will require lining to be strong enough for horses.

In addition, exterior walls and partitions can be lined with rubber to minimise injury to the legs of horses that kick out at the walls. However, this will add a lot of expense and should not be necessary. If a horse is kicking the walls, this is usually due to frustration at being over confined. In this case, the horse is stressed and the management of that horse needs careful scrutiny and changes need to be instigated.

Concrete walls, have good insulation properties and are therefore cooler in hot weather.

Stable windows

Glass or Perspex windows (or openings of any kind) increase ventilation in a stable complex, but are not as necessary in individual stables (e.g. a row of stables that open to the outside anyway). The top door of such stables should *always* be open, but if the top doors are closed on a regular basis for whatever reason then there should be an opening of some kind to let in light (and air).

If windows are situated within individual stables, they need to be barred and/or placed high up to prevent breakage and subsequent injury to horses. This renders them almost useless to a horse.

If windows are situated within individual stables, they need to be barred and/or placed high up to prevent breakage and subsequent injury to horses. This renders them almost useless to a horse.

Perspex windows can by all means be placed high up at strategic intervals in a larger building such as a stable complex in order to let light in, but, rather than equipping each individual stable with a traditional looking but frustratingly useless (to a horse) window, **you are better off spending money on either:**

- An opening at the back of each stable with shutters that can be left open most of the time, allowing the horse to put their head to the outside. In this case, they should be approximately 1.2m (3.9ft) square and 1.3m to 1.5m (4.2 - 4.9ft) from the ground. This arrangement is common in colder European countries and should become more common in hotter climates.

- An even better option is to have a stable door at the back of each stable in addition to the one that opens to the breezeway at the front of the stable. These doors should preferably lead to surfaced holding yards on the outside of each stable. This arrangement greatly improves the situation for a horse.

- Even if there is no outside surfaced holding yard, this door will be useful for ventilation *and* access both for every day and in the event of an emergency such as flood or fire.

Openings increase ventilation in a stable complex.

Stable doors

Individual stable doors should either open outwards (not inwards) or slide. Sliding doors should be fitted to the *inside* of the stable to prevent the possibility of a horse trapping a leg in the gap between the door and wall of an outside-hung door. Better still is for the door to slide into a front and back casing.

A door for horses that are 15hh plus should be at least 1.2m (3.9ft) wide and 1.4m (4.6ft) high – allowing horses to put their head over it easily. There should be 2.4m (7.9ft) of head clearance from the ground to any framing above the door so that a horse does not bang the top of their head when standing and looking over the door.

Some stables have barred or meshed top doors that prevent a horse from putting their head over the door, the idea being that it prevents horses from performing the 'vice' called weaving. However, even if it prevents the horse from weaving (sometimes the horse still weaves anyway within the stable), it does nothing about the cause and actually increases the stress levels in the horse because it reduces movement and view.

These stables have bars that can be put up or let down. It is highly questionable whether these are necessary.

Be aware that a horse that attempts to carry out a stereotypic behaviour such as weaving is demonstrating stress, and steps must be taken to improve the living conditions of that horse; as already mentioned, preventing the behaviour only adds to the stress. Horses do not learn these behaviours from each other as is commonly thought but, once developed, learned stereotypic behaviours can be hard to reduce, even if the source of stress is removed. If this or any other 'stable vices' are a concern for you and your horse, you should learn about what causes these behaviours to avoid increasing the stress levels on your horses.

In buildings where the general public are allowed access, there is a case for preventing horses from putting their head over the door. In this case, this should always be compensated for with an opening of some sort at the *back* of the stable. A stable that is fully enclosed is actually no better than a cage and, if this is where a horse spends most of their time, then this is totally unacceptable and is a welfare issue.

Roller doors like those on a modern garage can be fitted to the breezeway ends of a stable complex, or twin sliding doors can be used. Large swing doors are generally not a good idea, as these are hard to handle if the wind catches them.

These large doors open inwards so are not as likely to be affected by strong winds as large doors that open outwards.

Stable floors

The hard flooring of any area where horses are kept should be safe to use, have low dust properties, be level and easy to keep clean. Floors in a breezeway and individual stables can either slope slightly for drainage, or be flat and rely on bedding/sweeping for getting rid of moisture. A slight slope in a concreted stable is useful so that most of the moisture is soaked up by bedding, however the floor can be hosed from time to time when completely cleaned out. Underground drainage inside individual stables is usually too problematic to operate. Due to the nature of bedding and horse waste, underground drains tend to block up. Any open drainage channels in stables must lead to a septic tank system or water purifying system, rather than being allowed to run off and possibly enter a waterway.

The hard flooring of any area where horses are kept should be safe to use, have low dust properties, be level and easy to keep clean. These rubber pavers, laid in a breezeway, fit the bill.

A stable complex can either be concreted right through, including the individual stables, just in specific areas such as the breezeway and tack/feed/wash rooms, or not at all.

The advantages of concrete are that:

- It prevents seepage.

- It is easy to keep clean.

- It has low dust properties.

- It has a neat and tidy finish.

- It is relatively safe for people to walk on (e.g. they are less likely to trip while leading a horse).

The disadvantages are that:

- It is relatively expensive to install.

- It can cause slipping injuries and when people or horses do fall it is a hard surface to land on.

- It is cold underfoot. Both for people who are working on it for long periods of time and for horses that have to stand on it for long periods of time. In the stables, large amounts bedding will improve the properties of concrete however this takes more time to 'muck out'.

- It can be slippery for horses in particular; this problem can however be reduced by making sure that it is not finished with a smooth surface. If you are using a contractor who is not experienced with horses, you will need to make sure they understand what you want before they start laying and smoothing the concrete. They also need to understand that horses are heavy animals; concrete laid for horses needs to be at least 100mm (4 inches) thick.

Concrete laid for horses needs to have a rough surface.

Other options for the floors of stables besides concrete include limestone, compacted earth, sand or bricks laid on sand. The advantages of earth and sand are that they are cheap, safer for horses to move around and lie down on, and they require less bedding than concrete. The disadvantage is that they allow seepage of urine and manure; possibly into underground water. They can also smell and require periodic maintenance. Limestone has the advantage that it absorbs and neutralises ammonia. Bricks on sand are cheap (if the bricks are second-hand/recycled), but also allow seepage, can smell and can be slippery for shod horses (therefore need large amounts of bedding like concrete).

Rubber can be used for flooring in conjunction with other surfaces. Rubber mats can be laid on top of any surface to improve it. For example, rubber laid over hard floor surfaces such as compacted soil/lime or concrete also reduces stress on the horse's legs and reduces dust.

Mats can be purchased that 'lock' together, or large mats can be sealed between the gaps and at the edges so there is no seepage into the ground water and no smell. There are now even commercially made 'mattress' style rubber floors for stables. There are many options available for rubber flooring in stables, so shop around.

Even though rubber matting/flooring has an initial high cost, it has many positive features:

- A stable with a rubber floor requires less bedding, bedding is only required to soak up waste, encourage urination and to provide a bed. This means that there is less waste to get rid of and the stable can be cleaned out more quickly. It is a misconception though that no bedding is needed; a horse is unwilling to urinate on a hard surface. Also, horses that live in stables with rubber floors but no bedding become very dirty because they end up having to lay in their own urine and manure.

- The softer surface of rubber reduces shock to the horse legs and provides an even surface for a horse to stand on. This can reduce stress on a horse's legs considerably.

- Rubber is far safer as horses do not slip as easily as on concrete and for this reason it is a good idea to have rubber on concrete walkways as well. Keep in mind though that some rubber can be very slippery when wet. Use only products that are designed or approved for horses.

- It prevents horses digging holes in the stable floor. However, a horse that digs is stressed and needs adjustments to its management routine e.g. more hay and more turnout.

- The whole stable environment is less dusty due to less bedding. This advantage can be further enhanced by using the more expensive low dust varieties of bedding however, as less is needed, the overall cost is reduced.

- The rubber mats are usually made from recycled tyres, therefore rubber is a good choice for environmentally aware people.

The whole stable environment is less dusty due to less bedding. It is a misconception though that no bedding is needed.

Stable fittings

Individual stables' holding yards can be fitted with automatic drinkers and even automatic feeders. A swing out feeder can be fitted to the front wall of a stable so that a horse can be fed safely and quickly from outside. Rug racks, tie rings and hay racks are also considerations.

A swing out feeder can be fitted to the front wall of a stable so that a horse can be fed safely and quickly from outside.

There are pros and cons associated with automatic drinkers that should be considered if you are planning to fit them.

The case against having them:

- The initial expense; they cost a lot more than buckets!

- They require frequent checking for malfunction, which may mean that they stop working or even that they *don't stop* working, causing floods.

- It is difficult to monitor a horse's water intake, which is essential if a horse is ill, although it is possible to buy models that monitor consumption (more expensive).

- Some horses enjoy playing with them and can either continuously damage them, or cause floods; this is not surprising considering how unstimulating a stable environment can be for a horse.

The case for having them:

- They save on carrying heavy buckets of water.

- They cannot be tipped over by a horse.

- They don't run out as long as they do not malfunction.

Therefore, automatic drinkers can save labour and time, but they are not completely maintenance free. You may feel that the initial cost of installing them can be recouped by reduced labour costs on a commercial horse property, but not necessarily so on a non-commercial horse property. If you do decide to fit them, remember that they must be cleaned regularly (every day) as they can collect manure and feed. Even if manure/feed does not get into a drinker, it still needs to be freshened up daily due to the poor air quality in a stable. Without cleaning, a 'scum' settles on the surface of the water, making it unpalatable to a horse. Also remember that the pipes to drinkers can freeze in cold weather therefore they must be properly lagged.

Automatic feeders are another expensive fitting, but they do have their advantages for horses (usually competition horses) that are being fed concentrates, because they can then be fed smaller amounts more frequently – including through the night – which is safer than feeding large amounts less often.

A rug rack can be fitted to the wall next to the door on the outside of a stable so that rugs can be hung somewhere when not on the horse. The idea being that the rug is then out of harm's way. Horses that continually pull rugs off a door or rack and chew them, stomp on them etc. need something to do with their mouth. Horses need to be turned out for as much time as possible, with other horses, to graze and play. Increasing the amount of hay in the diet (more low energy hay, less high energy concentrate feed) are also remedies that should be considered.

A tie ring can be attached to the inside of a stable, but it may not be necessary if you are going to use specific tying areas or the outside of the stable for grooming and tacking (which is preferable). A tie ring must be very securely fastened and the horse tied to a 'weak link' of string between the tie ring and the lead rope. It is possible for a horse to pull a tie ring out of a wall when 'pulling back' and injure a person or themselves in the process. A common injury in this situation includes head injury/eye injury for people and horses. The string is a deliberate weak point and will break before either the head collar or the tie ring. It is also easier to cut in an emergency situation.

A hay rack inside an individual stable reduces wastage due to trampling; however feeding hay on the ground allows a horse to eat with their head in the natural grazing position. This is very important for horse health; in a natural situation, a horse has their head in the grazing position for at least 12 hours a day, helping to keep the airways and sinuses drained and healthy. The bones, joints and ligaments of a horse's head and neck have evolved to work effectively in this position. When a horse spends a significant amount of time feeding with their head at body height or higher, this can lead to sinus/airway problems over time. Also, a horse has to twist their neck to eat out of a hay rack- this is an unnatural feeding position for a horse.

Feeding hay on the ground allows a horse to eat with their head in the natural grazing position.

Other facilities

The horse facilities might include other areas such as a horse wash, veterinary area, hoof trimmer/farrier area, tack storage area, feed storage area, horse tie-up/tack up areas, laundry, kitchenette, toilet/showers etc. These areas can either be incorporated into a stable building complex or stand alone.

You may also need to build other buildings for storage of large amounts of hay, a workshop and for machinery storage etc.

Horse wash/veterinary/hoof care areas

A horse wash area can be indoors or outdoors. It is not imperative to have one; horses can be simply sponged off after work, but a horse wash can be a very useful addition on a horse property.

If you are building a horse facility and space allows, a horse wash can be situated within it or nearby, for example it could be situated on one of the outside walls so that water can be plumbed from the building to the horse wash easily.

On a small, non-commercial horse property, it makes sense to have a horse wash area that is also a veterinary area *and* a hoof care area. A larger horse property may warrant separate areas for these activities. Think about how much these areas will be used before deciding whether to combine them or whether to have them separate.

If you are combining them, it may be a better option to integrate this area into the main facilities building so that it is under a roof. Alternatively, if you are building this combined area on the outside of the main building, there will need to be protection from the elements. A simple horse wash is usually fine without a roof, but if people are tending to a horse, they cannot be expected to have to work outside in inclement or hot weather.

A wash bay/veterinary/hoof care area is usually the about the same size as a standard individual stable (3.6m x 3.6m - 12ft x 12ft). The flooring should either be 'rough' textured rubber (smooth rubber gets very slippery when wet) or 'rough' finished/patterned concrete. A novel idea for the floor of an outdoor horse wash is to have a layered gravel base, starting with larger rocks and finishing with smaller, rounder stones, which reduces slipping. If set up in the correct place this will also help to filter the water before allowing it to irrigate a pasture (you will still need to use biodegradable shampoos etc.).

The water hose for a horse wash can be raised on a swinging arm so that it is possible to wash both sides of a horse without the risk of it becoming tangled in a horse's legs. A spray attachment on the end of the hose reduces water consumption and makes it easier to use.

This outdoor horse wash has a gravel base.

Shelves for storing horse washing products and washing tools can be fitted in this area. They need to either be recessed or situated in a corner to reduce the chance of a horse banging in to them. It may be useful to have a sink fitted in this area too for washing horse gear. A heavy duty washing machine near this area could be useful if there is a lot of equipment to keep clean.

Hot water is another bonus. A domestic hot water system can be fitted. With an indoor horse wash, an infra-red heater is useful, but not essential, especially in colder climates. Keep in mind that solar is now a serious alternative for electric.

Water from a horse wash should drain into either a septic or grey water system. On a small horse property, the domestic system may be able to cope with the extra water from the horse wash, depending on how many horses are washed and how often. You will need to check the capacity so that you do not overload it.

Tack Storage

A tack room can either be part of a stable complex or in a separate building, but it should be close to the tacking area. Storage for horse gear needs to be secure (tack is prone to theft) and dry. Leather in particular becomes mouldy very quickly if kept in a damp environment. You need to manage rodents as well, as they can also damage tack.

A tack room within a stable complex can either be a simple area for storing tack, or a more elaborate area including a small kitchen/drinks facilities, heating, laundry and phone (office).

If space is a problem, a combined feed and tack room is an option but keep in mind that your tack will tend to get dusty (from the feed).

Hang tack on either purpose built or improvised hangers and racks for ease of use and some protection against damp and rodents and insects. Tack and rugs that are not in current use can be stored in plastic storage containers such as those that can be purchased cheaply from variety stores. This keeps spare gear dry and protects it from vermin and insects until needed. If you live in a particularly humid/damp environment, you can put moisture traps inside each box.

Storage for horse gear needs to be secure (tack is prone to theft) and dry.

Feed storage

It is convenient to store feed close to where it is needed; this will mean either having a feed room as part of a stable complex, or in a separate building near the stables or surfaced holding yard/s.

It is a good idea to have a smooth concrete floor in a feed room so that it can easily be kept clean by sweeping and occasional hosing. It is vital that rodents do not get into feed due to the associated dangers; vermin soil feed and also carry botulism. Horses are very susceptible to botulism poisoning. The presence of vermin also encourages snakes which can be a problem in some countries.

Feed can be stored in numerous kinds of containers, including purpose built feed bins, but these can be expensive and rather large. Clean, recycled drums made from plastic or steel can also be used. These are often sold inexpensively at animal feed stores. Old chest freezers that have had the clasps removed so that children cannot get trapped inside, also make good feed storage containers. The most important considerations for feed storage containers are that they are vermin proof and waterproof.

The amount of storage space required will depend on the number of horses to be fed, their feed requirements and how often the feed supply is to be replenished. Feed bins that are too small will need to be refilled every few days; too large means the feed may become stale before you use it. Remember that each feed bin should be emptied completely before being replenished so that feed at the bottom of the bin does not become stale and unpalatable. Otherwise, when refilling, remove the old feed, put in the fresh feed and then place the old feed back on the top (as long as it is still usable).

Feed can be stored in numerous kinds of containers, including purpose built feed bins.

A useful addition for a feed room that is within a larger building is to have an additional external door that has a large opening for ease of access when feed is delivered. A garage style roller door works well in this case.

Hay storage should not be close to stables for several reasons; large amounts of hay are a fire risk (both from self-combustion and external sources) and hay can also create a dusty environment in enclosed areas.

If possible, plan to build an additional hay storage shed that can be used for larger amounts of hay away from the other horse facilities, and the house (see the section **Other buildings**). It is better to bring only the amount needed for each day/feed into the stables at a time. So, for that reason, somewhere to store small amounts of hay near or in the stables or surfaced holding yards will be useful.

Keep in mind that is very useful to be able to store larger amounts of hay. This means that you will be able to buy larger amounts when it is available at a good price. Haymakers are usually happy to sell hay as soon as it is made rather than put it in their own sheds and move it to yours later, because this results in double handling.

Tying and tacking up areas

Tying areas with a smooth non slip floor are required for grooming and tacking horses and for dental work, veterinary work and hoof care etc. Whenever possible, horses should not be groomed inside a stable because of the dust factor (both for the sake of the person grooming and the horse). Grooming releases lots of hair and more importantly dander (dead skin), which will either be inhaled immediately by the occupants of the stable, or will fall to the ground to be inhaled later when the horse lies down. Aim to locate tying areas both outside and inside the building, and use the inside ones only when necessary in bad weather.

These tying areas need to be reasonably near the tack room for convenience. They should also be situated in a place where a tied horse will still be able to see other horses. A possible tying area is under a veranda on the outside of a block of stables; this gives shelter and shade while at the same time reducing dust inside the building. A veranda for this purpose needs to be approximately 3.6m (12ft) wide, as any narrower does not give adequate protection from the elements or room for a horse to stand.

Tying areas with a smooth non slip floor are required for grooming and tacking horses.

Tie rails (hitching rails) are another potential source of danger; *any* horse can 'pull back' when tied, even a horse that is very quiet and has never 'pulled back' could be stung on the nose by a bee for example! The term 'pull back' is used to describe what happens when a horse has a panic reaction while being tied up.

The horse does not usually stop panicking until something breaks. This is usually the rope, the headcollar, or the tie rail itself. In this case, the whole tie rail may be pulled out of the ground or the rail may be pulled off if it is not attached properly to the posts.

Either way, there are often disastrous consequences when this happens. So make sure that a tie rail is very strong and well-constructed and always use a breakaway section (string) when tying (or do not use the rail at all if it is not secure). It is far more important that people or horses do not get injured than that a horse stays tied. A tying area should always be enclosed by a fence so that a loose horse cannot escape onto a road.

Human comforts

Areas such as toilets, showers and a kitchen area are all very useful additions to a stable complex. There is no end to what you can build of course and this will be largely dependent on your needs and whether the building is a commercial building or a private-use only building.

A kitchen area is a very useful addition to a stable complex.

Other buildings

A building such as a three sided or fully enclosed farm shed is very useful on a horse property for bulk hay storage. Large amounts of hay should always be stored well away from anywhere that animals or people live because of the fire risk. Hay can be dangerous for two reasons; it can spontaneously combust soon after being baled and before it is fully dry, a common cause of hay barn fires, and hay is also a general high fire risk because it can be set alight so easily by a stray cigarette etc.

A farm shed is also useful for storing machinery such as a tractor/ride on mower and its implements. If the building also has an enclosed room at one end, this can be used as a workshop and/or for storage of tack and feed if you do not have storage elsewhere.

A building such as a three sided or fully enclosed farm shed is very useful on a horse property for storage.

Power and plumbing

The stable complex, other buildings and outside areas should be well lit for safety and convenience. Each individual stable should have its own light. Light and light fittings should be fitted well out of reach of horses and there must be a safety cut out switch fitted to the power supply. Halogen, incandescent or florescent lights can be used in individual stables; halogen lights last longer and are brighter than incandescent and tend to be more reliable than florescent.

Have lights fitted to the outside of buildings so that it is possible to walk safely from the house to the buildings at night.

A more ecologically friendly alternative is to use solar power for the horse facilities. This is becoming cheaper to install and, once set up, is very cheap to run. Using solar also gives you more flexibility when positioning facilities, as it is not necessary to be near the main power source.

You may also wish to have power points installed to your facilities so that you can clip horses/boil water etc. Power points can be fitted inside and outside buildings using heavy duty all-weather sockets. All electrical work should be done by a qualified electrician.

If you want to have running water to horse facilities such as stables, surfaced holding yards, the horse wash, automatic drinkers and taps, you will need to contract a plumber. At least one tap inside and one outside a stable complex is useful. If they do not have automatic drinkers, taps near the surfaced holding yards, will reduce the labour when carrying buckets of water.

Plan your horse facilities well and they will give you years of good service. Horse facilities do not need to be elaborate or particularly expensive, but they do need to be functional.

—

See the section *Horse facility positioning* for information about the *positioning* of horse facilities.

—

Manure storage

Provision should be made for proper handling and storage of any collected manure, along with a plan for its effective utilisation.

—

You need to decide where you will place an area dedicated to manure storage - see the section *Manure management planning*.

—

See *The Equicentral System Series Book 2 - Healthy Land, Healthy Pasture, Healthy Horses* for in depth information about manure management.

—

Chapter 2: Fences and gates

As well as helping to keep horses secured within an area *on* a property, fences protect certain areas from horses (such as waterways, or sensitive vegetation, or your garden) and maintain boundaries between a property and neighbours, or between a property and any public roads.

Good fences and gateways reduce the chance of accidents occurring to people, horses and third parties and therefore help you to sleep at night, knowing that your animals are safely confined on the property. Good fences and gateways are also aesthetically pleasing and increase the value of a property.

There is no one type of fence that is ideal for all situations, climates, types of horse or budgets. Traditionally, fences and gateways have evolved to make the best use of local and easily available materials. Stone and vegetation were and still are used in various regions throughout the world to good effect. Within reason, it is now possible to buy almost any type of fencing material wherever you live, meaning that horse property owners/managers now have plenty of choice.

Good fences and gateways are aesthetically pleasing and increase the value of a property.

Horse properties may have several different kinds of fencing throughout the property, as different areas often require different types of fencing. Using different fence types in different parts of a property is the best way of utilising the advantages of different fence types efficiently whilst avoiding the disadvantages of others; at the same time helping to keep expenses down so that you do not blow the budget. Therefore fencing used in a particular area depends on many factors including the paddock or surfaced holding yard size, the size and type of horse/s being confined, whether other types of livestock will be using the area and of course, the budget.

Horse properties may have several different kinds of fencing throughout the property, as different areas often require different types of fencing.

The management of the property also has a bearing on what type of fence can be used and good pasture management goes a long way in helping to avoid fencing issues. If grazing on the property is good, the horses will spend time grazing rather than looking for ways to get into the next area or trying to eat the grass on the other side of the fence (hence the saying - the grass is always greener on the other side of the fence!). Horses that have their heads down grazing do not tend to come into contact with the fences and gateways as much as lonely, hungry and bored horses do. If horses are kept in pairs or herds they are not as likely to spend time near fences and gateways looking for or at other horses, or even worse, playing with another horse over a fence or gateway. So, if horses have companionship, it vastly reduces the potential for fence injuries *and* land degradation issues such as tracking lines near a fence.

The *amount* of fencing required on a property also varies depending on how the land is managed. A balance must be struck between having enough areas to *rotate* the horses and any other grazing animals around the land so that pasture gets chance to rest, recuperate and regrow, and not having too many small areas which are difficult to manage and expensive to set up and maintain.

The grass is always greener on the other side of the fence! An example of a potentially lethal situation caused by management issues and therefore totally preventable.

The main considerations to think about when planning fences and gates are:

- **Cost** – the most expensive fences and gateways are not necessarily the safest or most effective; you can economise on some fences but not others.

- **Effectiveness** – the perimeter fence in particular is all that stands between your animals and the road. If your animals get out on to the road they can cause fatal accidents – to humans as well as themselves and you may be held liable for accidents caused by them.

- **Safety** – most equine vets will agree that after colic and laminitis, fence injuries are high on the list of serious dangers to a horse. Your main priority should be to reduce interaction between horses and fences. You can do this by having good pasture management e.g. so that horses do not put pressure on fences trying to get to better pasture and by acknowledging horse behaviour e.g. not allowing horses to play over a fence (either keep them together, or have double fences between horses - preferably the former).

- **Ease of construction** – whether you are constructing fences and gateways yourself or paying someone else, if they are time consuming to erect, they will be more expensive.

- **Aesthetics** – fences and gateways on a horse property should be reasonably pleasing to the eye. For example 'traditional' white painted fences are not necessarily the best, fences that blend with the landscape are often better (and easier to maintain).

The perimeter fence in particular is all that stands between your animals and the road.

Fence and gateway safety

Fences (and gateways) have the potential to be very dangerous for horses. Therefore, safety is an important consideration when designing fences and gateways for horses. Fences and gateways are a considerable expense on a horse property, but the most expensive options are not necessarily the safest or best. No fence or gateway is perfect, but good planning can keep accidents to a minimum.

Horses can and will injure themselves on any type of fence or gateway, even costly, traditional timber post and rail fencing, favoured by many because it is thought to be the safest option, can cause terrible injuries to horses if they attempt to jump or charge through it.

Keep in mind that fences and gateways are only dangerous when a horse actually comes into contact with them. Fences, gateways and horses do not mix, and the aim of good horse and land management should be to keep horses away from fences and gateways whenever possible.

There are certain things that you can do to reduce the risks:

- **Turn horses out together** to reduce the incidence of horses walking the fence lines or even challenging fences and gateways in an attempt to get to other horses. When horses are relaxed they will get their head down and graze rather than hang around fences risking injuring themselves. Remember that horses are herd animals and need companionship to thrive. By grazing horses in herds rather than singularly, you can also better utilise your available grazing by rotating pastures.

- **Add electric fence elements** to existing but undesirable fences and gateways to keep horses away from them and therefore away from this particular danger. The addition of electric to an existing fence also protects expensive fences and gateways (from chewing/rubbing etc.).

- **Ensure there are no projections** on fences and gateways. A common injury for people as well as horses is to be lacerated by sharp/hard projections on fences and gateways, so have a good look at your existing fences and gateways and 'risk assess' them.

- **Train your horses to stay calm** when their legs are caught up in something. This is a very important but often overlooked point. If a horse has learned to stand still rather than panic when caught up in a fence or gate, the damage will be minimal or even nil. It is a very different story with a horse that panics when they feel they are trapped. If you do not know how to do this, contact a professional horse person who specialises in handling and particularly desensitising/habituating horses to accept scary situations.

Adding electric fence elements to existing fences keeps horses away from them.

—

See the section **Gates and gateway safety in particular** for more information about the *safety* aspects of gates and gateways.

—

Fence visibility

Horses can and sometimes do run into fences (or gates), however it is usually due to panic rather than because they do not actually see them. A common scenario is when a horse is separated from other horses or when being chased by another horse. A panicked horse is capable of charging into *any* solid object as if they have not seen it.

When first turned into a new area, there is a greater risk of a horse not seeing a low visibility fence/gate, however once accustomed to a given area a horse generally knows exactly where any obstacles are.

Try to make sure a horse is in a calm state of mind before turning out into a new area so that the chances of injury are reduced. A horse that is too full of energy (e.g. too much confinement/too little exercise/too much high energy feed) is at a higher risk so it can be a good idea to turn such a horse out into a safer and more confined area, such as a training yard with high sides (within sight of other horses) until they are calmer (see Appendix: **Introducing horses to herd living** for more information).

The colour of a fence is not usually important, as a horse generally has good eyesight and even if a horse has sight problems, they tend to acclimatise to an area in time. Obviously, a fence with widely spaced posts and only a few wires will be harder to see than a fence that has more posts and some rails.

If it is thought necessary, an existing fence can be made more visible by adding a white electric tape, PVC sighting wire or a rail to the top of it. Alternatively, a fence can be made *temporarily* more visible by attaching plastic bucket lids to the top wire. Once a horse is aware of the fences, these can and should be removed as they are unsightly.

—

See the section **Horse facility positioning** for information about the *positioning* of fences and gateways.

—

Fence dimensions

For average sized horses (14hh-16hh), fences should be a minimum of 1.2m (almost 4ft) high, but consider having higher fences for the perimeter fence e.g. 1.5m (almost 5ft) when the fence is between horses and a public road; there is no harm in having them higher if you think it necessary. As well as being a better barrier, a taller fence prevents a horse from leaning over it to graze on the other side.

*There is no harm in having fences higher if you think it necessary (picture A).
As well as being a better barrier, a taller fence prevents a horse from leaning
over it to graze on the other side (picture B).*

The spacing between rails, wires or pipes should be close enough to prevent a horse from putting their head through the fence (approximately 20cm/8ins) *or* wide enough so that the head can easily be withdrawn if they do (approximately 50cm/20ins). With post and rail fences, a wire can be put between alternate rails to prevent horses from trying to put their head through.

Whenever possible, prevent horses from putting their heads through a fence, leaning over a fence to graze on the other side or even going near a fence. Aim to build fences (especially boundary fences) that horses cannot put their heads through to graze plants on the other side.

For many reasons, you want to prevent horses doing this:

- If it does not belong to you, the land on the other side could be contaminated with herbicides, pesticides etc. If this area is next to a road, the grass may be polluted by traffic, depending on the volume of traffic.

- When horses graze through or over a fence, they weaken the fence over time.

- And lastly, if a horse can get their head through a fence, small children or animals can also get in to the paddock through the fence. Dogs and children, whether they are your own or the neighbours, must be prevented from entering horse areas unless given permission (such as with your own children), because as well as not wishing anyone else's pets or children to get injured, you may be legally liable if they do.

You should aim to prevent your horses from going anywhere near fences, especially those that they can graze over or through. You can do this easily by using an offset electric fence or by making the fence higher and impenetrable e.g. using small aperture mesh or hedging (see the relevant sections).

The lowest element of a permanent fence (for horses) should be at least 30cm/12ins from the ground to reduce the risk of hoof and leg injuries, but can be higher, especially in the case of internal fences e.g. about 60cm/24ins; it depends on what other animals might be kept in the paddocks. If animals such as sheep, goats and miniature ponies etc. are going to be using the paddocks, small aperture mesh can be incorporated into the bottom half of the fence.

When horses (or donkeys!) graze through or over a fence, they weaken the fence over time.

Electric fencing on a stand-off/outrigger can then be used near the top to keep the horses away from the fence. For cows, a lower strand of electric is usually sufficient (around 75cm/30ins). This can be in addition to a higher strand of electric for the horses.

Temporary electric fences are usually set lower (a 'tread-in' post is approximately 75cm/30ins from the ground), because horses do not tend to go near them and are therefore unlikely to step over them or lean over them as they will with a non electrified fence.

The distance between posts depends on the type of fencing material used. With a post and rail fence, the posts are usually placed approximately 2.4m (about 8ft) apart but for a plain wire fence the posts can be placed further apart (4m/13ft) and several droppers used in between to hold the wires in place (droppers, also called battens and stays in some parts of the world, evenly space and support the wires in a fence - see the section **Droppers/battens/stays**).

The lowest element of a permanent fence should be at least 30cm/12ins from the ground to reduce the risk of hoof and leg injuries.

Fence types

Fence types tend to fall into one of the following four scenarios. **It is a good idea to put some thought into the following scenarios before deciding how to proceed:**

- **Expensive to install but with very little on-going maintenance requirements**. This includes pipe and steel fences, some (well-constructed and unpainted) timber fences, some mesh fences (such as commercial horse mesh) and PVC/vinyl fences. Stone walls/fences also fall into this category, but are usually too expensive unless you have a bottomless budget.

- **Inexpensive to install with some on-going maintenance requirements**. Electric fences fall into this category. Other types include hardwood posts with plain wire and some mesh fences (such as agricultural mesh). These require the addition of electric tape, braid or wire on a stand-off/outrigger order to keep horses away from them and reduce the maintenance factor (the electric stops horses from leaning over/pushing through the fence).

- **Expensive to install with high maintenance requirements**. This includes painted (usually white, sometimes green or black) timber fences. If you install this arrangement, you are signing up for long term high maintenance – especially if the fences are painted white. Cheaply constructed timber fences (without the addition of electric) are also very high maintenance - even if not painted - as horses lean over or push through them and snap the rails. In this case, cost saving is false economy indeed.

- **Inexpensive to install with high maintenance requirements**. This includes hedges – which take time to grow and do require regular maintenance; the horses may help a bit with this job depending on what plants make up the hedge, but are excellent in terms of safety, aesthetics, wildlife habitat and so on. They may need the addition of a solid type of fence (such as an electric fence) to make them secure enough. Certainly until they mature, but even then, there are not many types of vegetation that are capable of confining a determined horse without some extra help.

There are many different types of fence available, and different types of fence may be required in different areas on a property. Also, different fence types can be combined *within* a fence - for example timber posts with a single timber top rail (or round) and plain wires below the top rail. This gives the advantages of strength *and* visibility, being more aesthetically pleasing and more visible than a plain wire fence, but not as expensive as an all timber fence.

Different fence types can be combined within a fence.

Hedges

Hedges, along with stone walls, would have to be one of the most ancient fence types. They are common in most countries of the world and make good fences for many reasons.

The advantages of hedges include:

- They are relatively inexpensive.

- They provide a physical barrier which looks solid to a horse (especially to a horse travelling at speed) therefore in many cases they are the safest type of fence. If a horse actually crashes into a hedge, they will not usually receive the same level of injuries that they would if they crashed into a solid wood or steel fence for example.

They provide a physical barrier which looks solid to a horse (especially to a horse travelling at speed) therefore in many cases they are the safest type of fence.

- If they are made up of native species of plants, they provide habitat for wildlife which in turn help to control pest insects (flies, midges and mosquitoes etc.).

- They can provide some additional fodder depending on the plant type.

- See *Chapter 4: Horses and vegetation* in *The Equicentral System Series Book 2 - Healthy Land, Healthy Pasture, Healthy Horses* for more information about providing habitat for pest eating wildlife and the concept of fodder trees.

- They are aesthetically pleasing.

- They also provide a natural windbreak and some shade depending on the height. See the section *Windbreaks and firebreaks* in *Vegetation planning*.

- The height can be adjusted and a high hedge does not cost any more than a low hedge.

The disadvantages of hedges include:

- They take time to grow, but electric fencing can be used while they are growing. Once they have grown, electric fencing can continue to be used for additional security and to protect them from horses and other grazing animals.
- On-going maintenance may be required (trimming) but this should be worth it for the many benefits that hedges give.

If you are planning on planting hedges, you will need to do some research about which plants would make a good safe hedge in your locality.

Timber fences

One of the more expensive types of fence is a full timber fence. There are a few different types of timber fence, ranging from the very neat square cut posts and flat rails (a 'post and rail' fence), to more a rustic looking fence made from 'rough' cut timber. This type of timber fence is erected using no fixings – the rails are simply slotted into the posts. This traditional looking type of timber fence is usually more expensive because it is more labour intensive to erect.

A rustic looking timber fence made from 'rough' cut timber.

A 'post and rail' fence consists of timber posts and up to five rails depending on the height. When there are less than three rails, the gaps may be made up with plain wires or small aperture mesh. Timber fences are expensive because of the timber rails and the higher number of posts that are needed (on a wire fence, posts can be further apart and droppers used in between to reduce the number of posts). These fences are popular because they give the traditional look that is associated with horse properties, however they tend to fall into scenario three listed above - expensive to install and they can be high maintenance.

A post and rail fence can be painted (usually white, green or black) or stained black or brown if treated with oil. Old sump oil, usually free from mechanic workshops, can be used. White painted fences make a louder statement when new, but quickly start to look unkempt unless regularly maintained. Keep in mind that if you paint fences, you will need to carry this out for the life of the fence on a reasonably regular basis. You may also be devaluing a property, as this may put a possible buyer off if you try to sell! Oil treated fences blend better with the landscape and require less maintenance and, if old sump oil is used, they are more environmentally friendly than painted fences.

For safety reasons, the boards of a post and rail fence should be placed on the inside of the posts, as this prevents horses from knocking the rails off if they hit them front on, or from knocking their hips or other body parts on the posts if they travel at speed along the fence. Of course this is not possible where one fence separates two paddocks, unless a second rail is added on the other side – more expense. If not, horses can be kept away from the fence with the addition of electric fencing (see the section *Electric fences*). It is a good idea to do this anyway because this will help to protect the fence from chewing/rubbing etc.

The rails can be 10cm-15cm (4ins–6ins) wide by 2cm-2.5cm (about 1 inch) thick. Obviously, wider, thicker rails are stronger and look better, but are more expensive. They should be bolted (never nailed) with nuts and bolts or coach bolts onto the posts. All protruding fastenings should be countersunk to reduce the chance of injuries.

Posts can be set 2.4m (8ft) apart at their centres. By buying the rails in 4.8m (16ft) lengths and alternating the joins between posts the fence will be stronger.

The advantages of timber fences include:

- They are a good visual barrier to a horse moving at speed.

- They look good (if well maintained).

- They can be built and maintained by a good 'handyman' once a few key points are understood e.g. no protrusions etc.

The disadvantages of timber fences include:

- They are susceptible to fire, rot and termites, although certain products such as hardwood timbers and 'treated pine' are much better than others at resisting rot and termites.

- Horses can and do chew them, but this can be prevented to a large extent by protecting them with electric fencing; however you need to factor this into the costs. Horses that persistently eat timber fences are telling you that they are desperate for more fibre in their diet and therefore you need to feed more hay.

The boards of a post and rail fence should be placed on the inside of the posts, as this prevents horses from knocking the rails off if they hit them front on.

- Although often regarded as the safest type of fence this is not necessarily the case, as they do not give if a horse runs into them and they can splinter causing injuries. In addition, the fixings can cause injuries if they come free in a collision situation.
- They are relatively expensive and they need to be maintained, although the addition of electric should reduce the maintenance factor.

In short, a timber fence is often used as a front perimeter fence for its strength and aesthetics with other types of fencing used elsewhere on the property. When timber fencing is used, it is a very good idea to also use electric fencing (in the form of an electric tape, braid or wire in a stand-off/outrigger) to increase its safety factor and its longevity.

White painted timber fences are very high maintenance. Think very carefully before erecting them as they can put a potential buyer off if you ever decide to sell the property.

Pipe and steel fences

The advantages of pipe and steel fences include:

- They are resistant to chewing by horses, fire, insects (e.g. termites) and rot.
- They do not require paint and they have little or no maintenance costs.
- They are versatile e.g. they can be used as a paddock fence or for surfaced holding yards and arenas/training yards.
- They are very strong and they can be easily seen.
- Even though they are not traditional to look at, they are neat and tidy.
- If you use second-hand steel pipe you will be recycling.

The disadvantages of pipe and steel fences include:

- They can be expensive to erect. Second-hand steel pipe can sometimes be found in scrap metal/second-hand building material yards; in this case it may be relatively cheap. New pipe is expensive.
- They do not yield if a horse hits them at speed.

Various thicknesses of pipe can be obtained with 40mm-200mm (1.5ins–8ins) being the most useful. The thicker gauges can be used as fence posts in conjunction with other types of fencing material such as plain steel wire. The pipe 'rails' must either be welded to the posts or fitted with pipe clamps. Steel rails should be set at the same distances as you would use for rails on a timber fence.

The costing of this type of fence depends on whether you are able to buy the material second-hand or not.

In some countries, it is possible to buy ready-made steel fencing either as separate components which are then welded together or in ready-made lengths. This type of fence is expensive, but very effective. It is probably over the top for internal paddock fences, unless you have a large budget, but they make an excellent perimeter fence or a surfaced holding yard fence.

A recycled pipe steel fence, in this example being used as a training yard fence.

In some countries, it is possible to buy ready-made steel fencing either as separate components which are then welded together or in ready-made lengths. This type of fence is expensive, but very effective. It is probably over the top for internal paddock fences, unless you have a large budget, but they make an excellent perimeter fence or a surfaced holding yard fence.

Stone fences/walls

In Europe, stone walls are common in certain areas. They are usually very old if not ancient, having been erected a long time ago without the use of mortar. Hence, in Britain the correct name for them is 'dry stone walls'. Erecting new ones is very expensive because they require a lot of stone, labour and skill.

They can be seen in many other parts of the 'new world' too, particularly in areas that had stone on the ground that need clearing and immigrants that had the skills. In these other countries they are sometimes called 'rock fences', 'stone fences' etc.

Dry stone walls are usually very old if not ancient.

The advantages of stone fences include:

- They are a good visual barrier to a horse moving at speed.
- They are durable; they last indefinitely with some maintenance - some are already centuries old.
- They are aesthetically pleasing.
- They are a strong physical barrier to horse.
- They usually do not require a high level of maintenance.

The disadvantages of stone fences include:

- They will not give if a horse runs into them.
- They require skill to maintain.

If they are already on your property, they should be maintained and cherished; however erecting new ones is not usually an option due to the expense and high level of skill needed to erect them properly.

Stone walls are beautiful to look at and very functional.

Mesh fences

Mesh is a popular choice of fencing. It ranges in safety depending on how it is made and the size of the aperture in the mesh. It is usually either spot welded, linked or woven with interlocking joints. Whatever the type of mesh, the holes need to be small enough to prevent a horse's hoof from going through (about 5cm/2ins or less). Some types of mesh fence require a tight plain wire along the top and bottom of the mesh to prevent the mesh from sagging.

Shod horses are more at risk with some types of mesh fences (and plain wire fences) because the mesh/wire can get caught between a hoof and a shoe (even if the foot does not go through). This can result in the shoe either being simply pulled off, or result in a more serious injury to the hoof if the shoe pulls part of the hoof with it.

The advantages of mesh fences include:

- They keep smaller animals in and intruders out.
- They prevent horses and other grazing animals from putting their head through a fence.

The disadvantages of mesh fences include:

- Certain types can be somewhat dangerous.
- They can be very expensive.

Two common types of mesh fences are 'ringlock' and 'dog mesh' fence. Both have large gaps that a hoof can easily fit through. Dog mesh fence is spot welded mesh which can snap, leaving sharp protrusions. It does not work well on undulating country, as it cannot flex and accommodate the dips and rises.

Mesh fences prevent horses and other grazing animals from putting their head through a fence, but some types of mesh have large gaps that a hoof can easily fit through.

Ringlock is wire linked rather than welded, so is able to accommodate undulations to some extent. Both are unsuitable for horses if the horses are able to get close to the fence, but can work well as a perimeter fence if used in conjunction with an electric fence placed several feet in from it. A less satisfactory option is to put an electric stand-off/outrigger on the mesh fence itself. It can also be useful as a garden fence to keep dogs and small children in the garden and out of a paddock.

'Chicken mesh' fencing is sometimes used as part of a horse fence, especially on the lower half of the fence to keep small animals in or out. The stronger variety has been used successfully in foal paddocks as foals can bounce into it without injuring themselves.

'Cyclone mesh' fencing is good in small areas due to its strength and expense, and for keeping 'hard to contain' animals in and persistent intruders out. A full height (about 3m/3.3yrds) cyclone fence will certainly prevent anything from leaving or entering the property, but it tends to look industrial and ugly. Cyclone mesh fence also comes in standard fence heights and can be fitted to a steel pipe or wooden rail at the top, with either steel pipe or a very strong wire along the bottom edge. Make sure the bottom of the fence is either fitted to the ground or is at least 30cm/12ins from the ground to reduce the chances of hoof and lower leg injuries.

There are various commercial horse mesh fences types now available. They are at the expensive end of the range, but are an excellent type of fence. These types of fence are often used on horse studs because of their many safety features. The main ones being that foals bounce off them if they hit them at speed. These meshes are usually fitted between posts with a top rail. The top rail is mainly for aesthetics and also acts as a 'sighter' for the horses. The mesh does not require a bottom rail or wire. Once erected, the maintenance factor depends on the type of posts and top rail that are used as the mesh itself usually requires no maintenance.

'Cyclone mesh' fencing tends to look industrial and ugly but it is a very effective barrier.

There are various commercial horse mesh fences types now available. They are at the expensive end of the range, but are an excellent type of fence.

PVC/vinyl fences

These fences are for people who want the traditional look of timber fences without the maintenance. PVC or similar material fencing comes in two main types. One type is comprised of PVC fence 'rails' moulded over strands of wire (picture A). These fence 'rails' are supplied in long lengths that can be strained up to various types of post. They can comprise the whole or part of a fence, (e.g. three 'rails' or just a top 'rail' with plain wires beneath), forming a strong and flexible barrier. It is usually possible to buy black/brown as well as white 'rails'.

Another type of PVC fence consists of panels that look similar (at a distance) to timber rails, with posts to match (picture B). With *some* brands of this type of fence there are issues with shattering in cold climates and depending on the construction this type of fence *may* not be as strong as the first type, however there are various brands so do your homework as some are very strong indeed. With the less strong types the addition of electric will help to keep horses away from it. Modern PVC/vinyl fences are low maintenance and have a longer life span than when they first came on the market due to having better UV resistance.

The advantages of PVC fences include:

- They have 'the look' of more traditional types of fence, without the maintenance.
- They are usually easier to erect than post and rail fences.
- They are relatively low maintenance.

The disadvantages of PVC fences include:

- They may not last as long as post and rail fences, however the reverse can be true because they now have improved UV resistance and unless a post and rail fence is made *very* well it will not last long anyway.
- They can be very expensive however there is a large range in price with the PVC rails moulded over wire being the less expensive of the two types.

Wire fences

There are several types of wire available such as high tensile galvanised steel wire, soft galvanised steel wire, plastic coated steel wire, plain plastic wire and barbed wire.

Plain wire fences are very popular fences for horses in countries such as Australia, Canada and America where larger distances have to be covered, but are less popular in the UK.

Plastic coated steel wire.

Wire combines well with other fence materials to reduce the overall cost of a fence. For example, you could have wood posts and one wood top rail, combined with wires below.

High-tensile galvanised steel wire can be strained very tightly and can withstand a lot of pressure. The problem with high tensile wire is that if it does break, it can cause injuries as it snaps back and recoils. This depends on how the wires are fastened to the posts because, in some cases, the fasteners will absorb and slow some of the recoil.

Soft galvanised steel wire is more expensive than high tensile wire and requires more frequent re-straining as it tends to loosen more quickly. However, some horse people prefer it to high tensile wire because if it breaks, it does not attempt to recoil. Also, soft galvanised steel wire is usually thicker and so does not cut into a horse as easily.

Plastic coated steel 'sighter' wire is used for its strength and visibility; it can be strained tight and produces a good looking fence, however it is expensive compared to plain wire. It is also possible to buy plastic 'wire' that is able to be electrified due to having conductive strands built into it.

Barbed wire is not a good fence material for horses, it has many disadvantages on a property including:

- It does not prevent a horse from leaning on it or pushing through it; only electric will do that.

- Cows also tend to ignore it. If you are keeping cows, they can usually be kept in with plain wire combined with electric fencing.

- It causes a lot of damage to animals when they come into contact with it at speed e.g. if they run into it or kick out at it.

- It can also be damaging to wildlife e.g. flying/gliding animals such as bats and sugar gliders (Australia) can get hung up in it. It can prevent other wildlife species from moving around the landscape.

- It is expensive and frustrating to handle.

If you do have barbed wire on your property and it cannot be removed for whatever reason, there are some points to keep in mind:

- It must **never** be electrified.

- It is at its most dangerous when not strained tightly.

It can be made *safer* by installing a simple electric fence several metres to the inside of it. Trees and bushes can be planted in this gap, creating a wildlife corridor - remember to remove at least the bottom stand of barbed wire to allow certain sized wild animals to pass through. This turns an unsafe fence into a safer fence with many other benefits.

The advantages of wire fences include:

- They are relatively inexpensive.

- They are relatively low maintenance.

- They are reasonably safe if built properly.

The disadvantages of wire fences include:

- Some types can be dangerous

- They are relatively less 'aesthetically pleasing' compared to certain other types of fence.

It must be remembered that a wire fence is more dangerous if it is loose, therefore strainers should be built in so that the wires can be tightened regularly and easily.

*Barbed wire **does not** prevent a horse from leaning on it or pushing through it; only electric will do that.*

Electric fences

Electric fencing would have to be the most versatile form of fencing, and can also be a very safe type of fencing. The use of electric fencing means that horses are much less likely to touch a fence, let alone lean on it, trap their head or leg in it, rub on it or push through it. This in turn means that there is less chance of a horse damaging either themselves or the fence.

Electric fencing can be either permanent or temporary, make up the entire fence or compliment other types of fence. Electric tape, braid or wire can be installed and maintained by any handy person, as it can be mended with scissors or small wire cutters and it can be strained by hand, it is relatively inexpensive and, as it becomes more and more popular with horse owners, electric fence manufacturers are responding by producing better looking fences to cater for this market. It is an ideal fence for the sustainable management of land because it allows you to quickly fence off new trees or areas of land that need protection from livestock for one reason or another.

In this case a simple and inexpensive to erect electric fence is allowing the property owner to plant and protect new trees.

Electric fencing can be used in many areas of a horse property with the exception of small holding yards, riding arenas and training yards. It can be used on the boundary/perimeter fence (and is indeed recommended for this purpose) when it is either a permanent type of electric fence (with permanent fence posts rather than temporary ones), or when used in conjunction with a solid type of fence.

Electric fencing therefore works best in the following situations:

- As internal fencing (either permanent or temporary).

- To make unsafe fences safer (e.g. certain types of mesh or wire fences).

- To protect certain types of fences (e.g. timber post and rail).

- To decrease the chance of a horse challenging the fence (e.g. a perimeter fence).

The advantages of electric fencing:

- It is extremely versatile.

- It is easy to set up and anyone can do this.

- It is relatively inexpensive.

- Fences can be curved easily.

- It works well if you are grazing cows on your land in addition to horses.

The disadvantages electric fencing:

- It can be shorted out by vegetation that grows high enough to touch it, or twigs and branches that fall across it. Regular checking will remedy this.

- In very dry weather, such as that experienced during a drought in Australia and the USA in particular, it can be difficult to get an earth.

- It can increase the fire risk in fire prone areas. Therefore, it should be turned of on high fire risk days especially if there is chance of plants touching it.

- It may be prohibited in certain urban areas (check with your local authority) When it is permitted in these areas, it is usually a requirement that it is clearly signed as such at frequent intervals along the fence.

*This is a situation that is **not** suitable for this type of electric fencing - as the sole fence between a horse and a road.*

Energisers

Electric fences require an energiser and this can be either a fixed mains operated unit, a fixed or portable solar/battery unit, or a portable battery unit that uses either rechargeable or non-rechargeable batteries. A mains operated unit will need to be situated near a power source, so it is usually preferable to site it inside a building for security. Solar/battery or battery units are both portable, which makes them very useful but also susceptible to theft. Energisers are available in various sizes so you need to work out how many kilometres/miles of fence you need to energise at a time. Most small horse properties can be adequately energised with one of the smaller units (5km/3m or less), depending on how many strands you plan to have electrified and whether you will be electrifying all of the land - all of the time, or just one paddock at a time.

A fixed solar unit.

Tapes, braids or wire

There are various options for electric fencing ranging from electric tapes of different thicknesses, electric braids (ropes) of different thicknesses, or the thinner types of galvanised steel wire. Galvanised wire conducts electricity better than tapes or braids and lasts longer, but is not as visible and, for this reason, it is best only used as a stand-off/outrigger on a permanent fence, rather than as a temporary electric fence (such as those used for strip grazing). Braids tend to last longer than tapes but tapes are the most visible option.

The efficiency of an electrified tape, braid or wire depends mainly on its ability to maintain voltage over the length of the fence. Any leakage of current in large enough amounts can make the fence ineffective.

The effectiveness of a single electric tape, braid or wire depends on the electrical circuit made when the horse comes in contact with the wire. The horse may not get a 'kick' from the fence if the soil is very dry, or if the earthing system at the energiser is insufficient.

The thicker tapes work best when threaded through carriers such as this one.

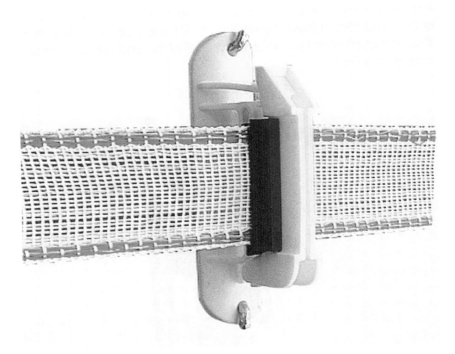

In drier areas, it may be necessary to incorporate an earth-return wire with a live wire to get satisfactory results. Nearly all permanent electric fences are designed with a number of live wires and a number of earth wires to achieve maximum effectiveness.

It is a good idea when designing your electric fence layout to incorporate a number of cut-out switches to isolate sections of fence for working on, or for fault finding; this way paddocks will be able to be switched off when not in use. There are many, usually cheap options, for you to use; consult your fencing supplier.

An example of a cut-out switch that can be incorporated into a fence so that sections (e.g. paddocks) can be switched off when not in use.

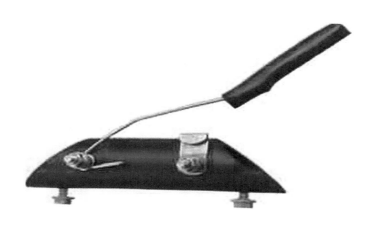

Where an electric fence comes to a steel or timber gate, an insulated electrified wire can either be run underground at the gateway or an electrified 'gate' can be used across the front of the solid gate to keep the horses back from it. This reduces the chance of head and hoof injuries from gates.

An electric fence tester is a must and, if you can afford it, a digital readout tester is far superior, as it will show you how much voltage is pulsing down your fence-line. This is especially important when tracing partial short circuits. Some testers have an in-built arrow system, which shows you in which direction the fault is from the tester. Others have a device for stopping and starting the electric fence from wherever you are along the fence so that you can either get through or repair the fence.

While it is true that electric fences are not generally dangerous for horses, you must be careful when introducing a horse to electric fencing for the first time. Horses quickly learn to respect an electric fence, but they first need time to learn about it. For example, never turn a group of inexperienced (to electric fencing) horses out together; they each need to learn about the fence while they do not have to cope with the other horses in the group.

Likewise, when a new horse that is not familiar with electric fences is brought on to a property, they should be put in a paddock alone with other horses in sight, but not directly on the other side of the electric fence, so that they can learn about electric fences before gradually being introduced to the other horses. Never hold a horse as they touch an electric fence for the first time – you can be badly injured doing this. It is better that a horse learns about electric fences without you in the way.

An electric fence tester is a must.

Most of the electric fence manufacturers have excellent websites and downloadable product brochures which detail how to set up electric fencing on your land.

It is possible to archive a neat and tidy, and effective, fence using inexpensive materials.

Electric stand-offs/outriggers

The cheapest and quickest method of making an existing fence horse-proof either to protect the horses from the fence or vice versa, is the addition of a single electrified tape, braid or wire, attached approximately 30-40cm (about 1ft) away from the fence by stand-offs/outriggers (picture A and B).

Stand-offs fasten to a fence so that an electrified tape, braid or wire can be passed through them. They can be various forms including wire loops, fibreglass rods, polypipe lengths, or hardwood brackets.

Some can be attached to the wires of the fence, while others can be drilled into, stapled or bolted to the fence posts. Metal stand-offs need the addition of an insulator to carry the electrified wire.

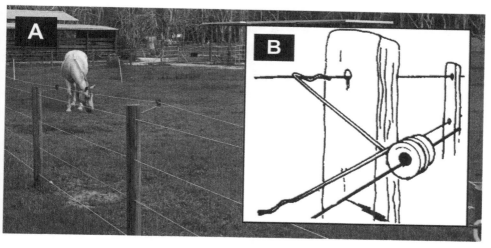

On an existing permanent fence, one electrified tape/braid/wire fitted near the top, level with or higher than the top of the fence is usually enough for horses. Sometimes it may work well to fit the electric stand-off directly *above* the fence, because this increases the height of the fence *and* it works for both sides of the fence at the same time. If other types of animals also graze the paddock, it may be necessary to fit a second electric element at an effective height for these animals - usually lower than that for horses, unless you have miniature ponies and Friesian cows!

Sometimes it may work well to fit the electric stand-off directly above the fence.

Novel and inexpensive stand-offs created by chopping lengths of polypipe. An issue to be aware of however is that a horse may inadvertently graze under such a large stand-off and bring their head up between the fence and the electric wire. A second stand-off fitted further down would make this situation safer. It would also make the fence more suitable for cattle.

Fence posts

Fence posts can be made from a variety of materials including timber (round, split, treated or untreated), 'polypipe', recycled plastic/composite materials, concrete or steel. Some fence posts are for electric fencing only, but some will suit electric or non-electrified fencing. Electric fence posts tend to fall into two categories; temporary/portable fencing posts or permanent fencing posts.

Tread-in posts

Tread-in posts are for temporary/portable electric fence systems and are commonly used when 'strip' or 'block' grazing a paddock. They are called tread-ins because they can be put in the ground with your foot. They are best used in conjunction with a *thin* electrified tape or thin electrified braid, either of which will usually snap if a horse charges into them. They should *not* be used with the thicker types of electric tapes/braids/wires (picture A); these are not as likely to snap and can result in a panicked horse running around with unbreakable electrified wires and tread-in posts that have lifted out of the ground trailing behind them. These posts are made of UV stabilised PVC (picture B) or steel with a PVC covered 'pig-tail' top loop (picture C).

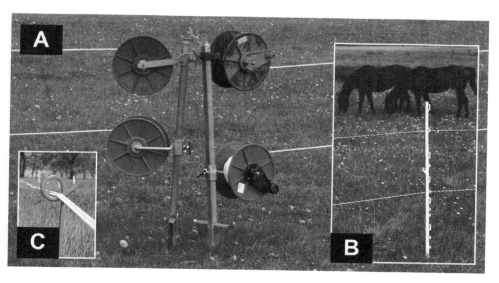

Fibreglass posts/rods

These are electric fencing posts that are self-insulated. Electric tape, braid or wire is usually attached to them with stainless steel clips (picture A). These posts can bend without breaking and are ultraviolet stabilised. They are available in a range of lengths with 10 mm (0.3ins) diameter posts for temporary electric fences and 13 mm (0.5ins) diameter posts for permanent electric fences.

Insultimber posts

In some areas/countries it is possible to purchase a product called 'Insultimber'. These are a dense, very long lasting eucalypt (Ironbark) product (picture B) and they are used for posts *and* droppers for electric fences. They are grown in Australia, but are also available in other countries. They have good insulation characteristics and do not need extra insulators attached when used in electric fencing systems. Hardwood posts are better than softwood posts in countries that have termites (white ants).

Other hardwood posts

Hardwood posts (split or whole) are longer lasting than soft wood posts. You should enquire as to what is the best type of hardwood and its life expectancy in your locality. Timber posts can be cut slanted or capped at the top to reduce rotting. Recycled railway ties (sleepers) can also be used as posts and they will last indefinitely due to the preservative they are treated with. They give a wide, flat area for fastening rails to which makes them ideal for post and rail fencing.

Softwood timber posts

Softwood posts (usually pine) are usually treated; otherwise they would not last long. They are more expensive than untreated wood, but generally outlast them by as much as four times. Any wooden post will burn in a hot grass or bushfire, regardless of preservative treatment. Untreated timber posts are also susceptible to rot and termites.

Composite/recycled/plastic posts

It is possible to buy hollow heavy duty plastic pipe (Polypipe) that has been cut to fence post length (sometimes recycled from use in the mining industry etc.) and recycled/composite material fence posts (available as other products as well). Both of these types of post particularly suit electric fencing due to their insulation properties.

A hole for wire can be drilled right through, thereby reducing the need for extra fixings. When using hollow Polypipe posts, remember to drill a large hole near the bottom so that rainwater can escape; other wise it could short out any electric wires threaded through the post.

The choice of such posts is likely to become more common as time goes on. They are strong, long lasting and resistant to acid, salt, wind, water and frost.

Steel posts ('star pickets')

Steel posts (galvanised or not) have various advantages and disadvantages on a horse property. They *can* be dangerous because it is possible for a person or a horse to be staked on them however, if they are used correctly, they can be invaluable. **They are:**

- Long lasting.
- Relatively inexpensive.
- Easy to put in by hand (with a hand help post rammer).
- Good in undulating and/or rocky country.
- Reasonably fire resistant.

They can be made much safer with the addition of caps or complete sleeves that are purpose made for these posts. Avoid using steel posts both in areas such as near an arena, where a rider could fall onto one, and in areas where horses can push each other onto one (e.g. a gateway). They work well when used as part of an electric fence system which keeps horses away from them. Plastic wires clips should be used when attaching wires to them rather than dragging the wires through the holes in the posts - this will prolong the life of the wires. To reduce fencing costs and increase safety, place the posts further apart and use droppers in the gaps between them.

Steel posts can be made much safer with the addition of caps or complete sleeves that are purpose made for these posts. They work well when used as part of an electric fence system which keeps horses away from them.

Droppers/battens/stays

Droppers reduce the total cost of a fence because they are less expensive to buy and fit than posts. They also make it more difficult for animals to push through a fence, although, if the fence is electrified, they will not do this anyway.

There are various options for fixing the wires/tapes to the droppers such as:

- **Holes** – the advantages are that once the wires have been threaded through the holes, they are not going to come out. They also save on buying clips for the wires. The disadvantages are that threading the wires through holes in droppers is time consuming and makes the job of fencing much harder. The resistance of each wire passing through one or more holes in droppers (depending on how many you have between the posts) creates more overall resistance. This may mean that the wires cannot be pulled through by hand. This dragging can also cause the protective galvanised surface on the wire to be damaged and therefore shorten the lifespan of the wire. If you are using electric wire/tapes, you should find it easier to strain the fence so the above points may not be a problem.

- **Slots** - the advantage is that the wires/tapes can be pressed into the slots *after* they have been strained (tightened) between the end posts. The disadvantage is that they are not as secure as holes or clips.

- **Clips** – the advantages are that, like slots, the wires/tapes can be attached *after* they have been strained (tightened) between the end posts. Clips allow a materials such as steel droppers to be used as part of an electric fence because they insulate the wires. The disadvantage is the extra expense (to buy the clips) and the time it takes to fasten them to the droppers – remember, you will probably be attaching at least two or three to each one.

A further choice is whether to buy full or partial length droppers. Full length droppers rest on the ground, giving the wires more support. Partial length droppers do not reach the ground; in this case they simply space the wires evenly. Apart from full length droppers costing more, the main consideration is usually whether the fence is electrified or not. Full length steel droppers for example will need clips to insulate the wires if the fence is electrified, whereas partial length droppers will not.

There are various options for fixing the wires/tapes to the droppers including holes and clips.

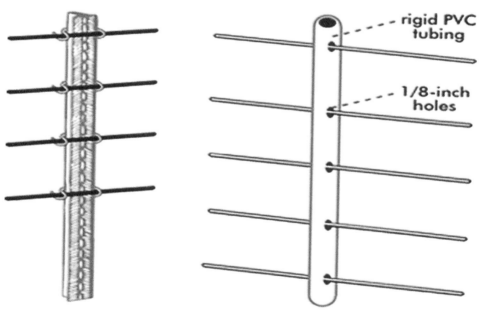

Droppers are usually made from one of the following materials:

- **Wood** - these can usually be purchased in various lengths as either plain board or with holes or slots pre-drilled. If they have holes pre-drilled, plastic clips can be fastened through the holes if necessary. It may be possible to buy certain *hardwood* droppers for electric fences that are self-insulating; hardwood is a better insulator than softwood like pine, for example.

- **Steel** - pre-formed light-gauge steel droppers in various lengths are commonly used as droppers and usually have slots. They cannot touch the ground if they are being used for an electric fence.

- **Plastic** – various options are available in plastic, either commercially made, or you can make your own from 2.5cm/1in 'polypipe'. Simply chop your required length and either drill holes right through them or cut grooves into them to hold the wires.

- **Recycled composite materials** - these are available as droppers but also as fence posts etc. Do a Google search for recycled composite droppers/fence posts and you will find various options.

Gates and gateways

Gates and gateway safety in particular

Gates and gateways are an area of high activity on a horse property and are potentially *the most dangerous section of a fence* for various reasons:

- **Horses tend to congregate in gateways**, waiting to be let in for feed or because that is where they get fed.

- **People and horses come into close contact** with the gate and its fastenings when passing through a gateway.

- **People and horses come into close contact** with each other in gateways, so many injuries occur to people in this area.

Common problems and solutions associated with gates and gateways include:

Problem - injuries when a horse becomes 'cornered' whilst being released into an area that already contains other horses. Sometimes a horse will try to jump a gate in this situation in an attempt to get away. This situation is also very dangerous for people if they are caught up in the melee.

Solutions - introduce horses carefully to other horses before turning them out together (see Appendix: *Introducing horses to herd living*). Avoid having a gate fitted in the corner of a paddock, aim for at least 3m to 6m (about 3 to 6yrds) in from the corner if possible. If the gate is already situated in a corner of a paddock, fit an electric tape or braid with an electric 'gate hook' diagonally across the corner (picture A), making sure this 'gate' becomes 'dead' once it is unhooked from the fence. This electric gate keeps horses away from the original gate when in the paddock and makes it easier and safer for both people and horses when putting a horse in or taking one out of the paddock. It also reduces the chance of a horse getting 'cornered' and attempting to jump the gate as a result. A more expensive, but even better option is to create a permanent second fence diagonally across the corner, or to create a square surfaced holding yard around the original gateway (picture B).

This then means that there is a surfaced holding yard in the corner of the paddock which is a very useful addition for various reasons. The new gate can be at the junction of one of the sides of the square – well into the paddock - which puts it into a far safer position.

Problem - injuries to the face and neck of a horse (sometimes fatal) when a horse gets their head or neck trapped between a gate and a gate post. Horses will rub their face or ears on anything given the opportunity. Similar injuries can occur where two gates meet. The gates cause a nutcracker action trapping a horse's head/neck or leg/hoof between them.

Solutions - make sure gates are fitted *flush* to posts so that there are no gaps for a horse to get any part of themselves into. Fit double gates *flush* to one another with a tight fastener at the top *and* bottom. Or keep horses right away from the gate by using an electric fence 'gate'.

Problem - injuries to hooves and legs if a horse gets itself trapped in the gate or between the post and the gate when 'pawing' at the gate; a very common injury.

Solutions - a gate should be meshed with small aperture mesh so that a horse cannot get a foot through it, or should have large enough gaps so that a horse can get a foot back out easily (the first option is by far the preferable). Alternatively, keep horses well away from the gate by using an electric 'gate' as mentioned above. Avoid feeding horses in a gateway or you will be encouraging them to 'hang out' in this dangerous area.

*This gateway is particularly dangerous because there are various gaps that a horse can get parts of themselves caught in **and** it has large aperture mesh.*

Problem - injuries to hooves and legs caused by gates with a flat strap of metal diagonally across the middle. A 'pawing' horse can get a hoof caught in the narrow area where the diagonal strap meets an upright metal part of the gate. The metal strap is particularly dangerous in this case because it has a relatively sharp edge on it. Any gate with a diagonal cross-member has the potential for a horse's hoof to become trapped in it as the hoof 'slides' down into the narrower section.

Solutions - make sure a gate has vertical parts rather than diagonal parts. Better still, mesh the gate so that a horse cannot get a leg through it.

Injuries to hooves and legs caused by gates with a flat strap of metal diagonally across the middle (picture A). Makes sure a gate has vertical parts rather than diagonal parts (picture B).

Problem - injuries caused by a horse jumping or attempting to jump a gate because they are trying to get towards another horse. Sometimes a horse will panic if a 'paddock mate' is taken out of sight and will attempt to jump a gate (or the fence) in order to get to them.

Solutions - if there is any chance that a horse might do this, the horse should be placed in a secure, high-sided surfaced holding yard rather than left behind in the paddock when the other horse is taken away. Spend time getting the horse in question used to the other horse being moved *gradually* away and eventually out of sight. This may take quite a while.

Problem - injuries to the body of a horse from a protruding gate fastener as a horse moves through a gate or slams into a gate while galloping around a paddock. A protruding gate fastener can also cause serious injuries to a person if they are pushed into the gate opening by a horse when leading a horse through a gate. Another potential injury can be caused if a horse gets the leg strap of a rug hooked up onto a protruding gate fastener while passing through a gate.

Solutions - make sure any gate fasteners do not protrude at all. This is sometimes difficult to archive because many of the gate fasteners that are sold (in horse/farm/feed stores) are dangerous for horses - even so called 'horse gate' fasteners. Gate fasteners can usually be made much safer by recessing them into the post out of harm's way. As gateposts are usually timber, this is possible in most cases. If not, fit the fastener so that it does not protrude into an area where a horse or handler could get caught on it. Alternatively, place the fastener on the other side of the post, in the fence line, and extend any chain on the fastener so that it still reaches.

*A protruding gate fastener (picture A) can cause serious injuries to horses **and** people. Make sure any gate fasteners do not protrude at all (Picture B).*

This kind of gate fastener is commonly sold to unsuspecting horse owners. It looks great when fastened (picture A) but is potentially lethal when open (picture B).

Problem - a horse can be trapped between the gate and the post when passing through a gate that is opened inwards (towards the horse). Horses tend to panic if they feel anything tighten suddenly around them; this can cause a very serious situation indeed. If the gate faster protrudes into the opening, the potential for injuries will be even worse. Traditional books on horse management often say that gates should only open inwards to prevent a horse from pushing a gate open and getting out of a paddock. This is not safe for people or horses. Never stand in the corner with a gate behind you that does not open outwards as you can become trapped by a horse that suddenly decides they want to get out. For example, if another horse comes up behind the horse and they panic and run forward, you will be trapped and crushed - you **must always** have a quick escape route for yourself.

Solutions - paddock and surfaced holding yard gates should swing freely and open both ways. Other solutions include those outlined above, for example fencing across the offending gateway to create a new safer gateway etc.

Problem - a gate can be lifted off its hinges and fall onto a horse when rubbing on it; horses particularly like to rub the base of their ears on anything handy.

Solutions - make sure the gate cannot be lifted off the hinges if rubbed by a horse. The hinges should have caps that prevent the gate from being lifted off. Most modern gates do but some older gates don't, so you should always check. Don't tie horses to gates or encourage them to hang around them by feeding them in this area.

*Make sure the gate **cannot** be lifted off its hinges.*

Problem - a horse plays with a gate fastener in order to open the gate; as well as creating the potential for accident if a horse gets out on the road, a horse can rip their lips depending on how the fastener is made (a common injury). 'Dog clips' and in particular spring loaded dog clips are particularly dangerous.

Solutions - use gate fasteners that are horse tamperproof. In some cases a padlock is best but only use if there is minimal chance of fire in which case the horses would not be able to be removed easily. If they *are* used the key should be kept nearby and in an easy to locate spot, alternatively the number of a combination style padlock should be displayed nearby in case of emergencies.

'Dog clips' (picture A) and in particular spring loaded dog clips (right of picture A) are particularly dangerous. In some cases a padlock is best (picture B).

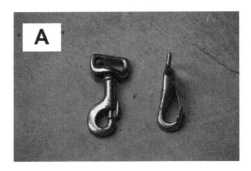

Gateway dimensions

A gateway should be at least 1.2m (4ft) wide for horses, but it is usually better to make a gate at least 3m (10ft) so that it can be used for machinery as well. If the gap is very wide, you may consider having two gates to fill the opening instead of one.

Be aware though that having two gates can also cause problems:

- They may both swing open when you only want one to open.
- They are not secure unless well fastened at both the top *and* bottom.
- Horses can get parts of their body such as their legs or head stuck between them, particularly if they are not fastened both top and bottom.

Alternatively, a more expensive but safer option is to have two gates (e.g. one large and one small) separated by a fence post.

Here there are two large gates (for vehicles) and one small one. The small one is too small for horses but fine for humans. If it was wide enough to lead or ride a horse through it would be a more versatile situation.

Gate types

Gates can be made of various materials with some of the most common being galvanised metal, timber and electric fence components.

Galvanised metal gates are the most common type; however some are better than others. They are generally mass produced for cattle (and other animal) properties rather than horse properties.

Gates (and gateways) have various issues to do with safety (see the section *Gates and gateway safety in particular*) in addition to the information in this section.

It is possible to buy, or have made, gates that are designed for horses e.g. they do not go down to ground level. This means that there is much less for a horse to get caught up in.

An example of a gate designed for horses, because it does not go down to ground level there is much less for a horse to get caught up in.

A 'gate' can be made from electric fence tape or braid and a plastic handle. There are also purpose made electric gates made from coiled wire and bungee rope on the market. An electric gate can be used as part of your internal fencing (e.g. the horses will not be out on a public road if they get out through it) or they can be used to enhance a solid gate. If you are using an electric tape gate, set it up so that it is not electrified when you unhook it to lead a horse through. This can save a potential accident if a horse accidentally steps on the 'gate' while being led through the opening. For this reason, make sure the fastener has a place to be hooked back to (and that keeps it taught) when opened (rather than leave it lying on the ground). Keep in mind that the period between when an electric fence gate is unhooked and then hooked up again is the most dangerous time.

A timber gate looks good, but usually costs more than a steel gate. They can sag over time if they are large and therefore heavy, so it is a good idea to rest the opening end on a stump or stone etc. when it is in the closed and open position.

Alternatively, the opening end can be fitted with a wheel to help support the gate and help with opening and closing.

This is a good galvanised gate for horses because it is meshed with small aperture mesh.

A timber gate looks good but they are heavy.

—

For information about **cattle grids** and horses see the section ***Property access***.

—

Chapter 3: Riding arenas and training yards

An all-weather surfaced area for training and exercising horses can be very useful, and many would say essential, on a horse property. It all depends on what you do with your horses. An all-weather surface can be various shapes and sizes, depending on your chosen discipline.

At the planning stage you need to think about the following factors:

- How often will you use this all-weather surface, once or twice a week, or will you use it on a daily basis for many hours a day?

- Will this area be used solely for riding and training or will it also be used as a surfaced holding yard? This is an important consideration if you have a limited budget and/or space is at a premium, as is the case with small horse properties (see the section *Can this area be multi-purpose?*).

There's no harm in dreaming! This beautiful indoor school built hundreds of years ago (in Sweden) is still in daily use.

- Are you planning on having a riding arena or a training yard - or both? You may be able to manage with just one area or you may need several, each fulfilling a different purpose.

- Is it to be outdoor or indoor? Will you start with an outdoor all-weather surface and put a roof on it later?
- Are you planning on having all the work carried out by a professional company or are you planning to do some or all of the work yourself (DIY)?

You will need to do some research before starting your project. As mentioned before, when planning and building any horse facilities, it is a good idea to talk to other people who have carried out similar projects and see what you can learn from them. If possible, also have a look their all-weather surface. Ask questions about the construction of it.

The sorts of questions to ask are:

- Did they construct the all-weather surface themselves or did they use a contractor?
- Have they had any problems with the base or surface?
- Have they changed or added anything to the surface since it was constructed?
- Would they like to change anything?
- Did they encounter any problems during construction?

The answers to these questions and any others you can think of will be invaluable in helping you with the construction of your own all-weather surface. Keep in mind though that weather conditions vary from area to area and materials from one quarry will not be identical to materials from another, therefore just because a particular base and surface has worked in one instance, it does not necessarily mean that it will work in another.

Many of the issues when planning and constructing an all-weather surface are the same for both a riding arena or a training yard, but some are different. At the planning stage, you need to think about what size and shape the all-weather surface will be and what the top surface will be made of. You also need to think about whether the all-weather surface will be fenced and, if so, what with. The above points are covered in the following sections. At the planning stage, you also need to think about the positioning of an all-weather surface. This is covered later.

The following information will help you to ask the right questions of any potential contractors and will help you to give contractors the right directions about what you really want. If you plan to DIY your all-weather surface, this information will help you to decide if that is the best way to proceed and if you do – what the important considerations are. This information is just a guide; you will also need advice from a local earthworks expert. Plan well and get as much qualified advice as possible before starting.

Ask other people that have an arena or training yard what they like and dislike about theirs and learn from their experiences.

Check with your local authority prior to the planning and construction of an all-weather surface, as there are often considerable earthworks involved that could impact on the local water catchment, on protected trees or have other environmental implications. Council regulations may limit use of the all-weather surface to 'personal use only' which may prohibit its use for commercial purposes such as coaching visiting riders etc. Better to find these things out before building. (see the section **Building permits**).

This roundyard has a permanent surface but is fenced with panels which means that the size and shape of the roundyard can be changed relatively easily.

Do you really need one?

If you are a professional coach or trainer or you train your horses regularly then a proper all-weather surface is likely to be essential for you. On the other hand many horse people think that they need an all-weather surface when in fact they can manage without one. For example, if you already transport your horse/s to lessons (some coaches will not travel) or to riding club facilities, you may not get enough use out of having an all-weather surface at home to warrant the expense. Or, if a neighbour already has an all-weather surface and is happy to let you use it (for free or for a fee), this may be a better (and cheaper) option than building one on your own land.

A privately owned all-weather surface is generally used for less than seven hours a week if the owner works off the property and only has time to ride/train their horse/s after work etc. On the other hand, such all-weather surfaced areas on a commercial property are often used for many hours a day and without them the business would probably not be viable. You need to decide how much expense can you justify, because a dedicated all-weather surface can be costly. Can you also justify large amount of space that this area will take up?

The positioning, size and shape and whether to put a fence around the area are all important considerations when designing your riding arena or training yard.

Work out the costs and benefits, taking into account how many hours a week you ride/train, although this figure may increase once you have a proper all-weather surface to ride/train on because adverse weather will be less of a factor, and weigh these up against not having one, in order to help you make your decision.

If you coach riders or train horses then a surface of some kind will probably be high on your list of necessities. Otherwise you may find it hard to justify the costs of installing a surface.

Can this area be multi-purpose?

If your budget is limited or space is a factor, it may be easier to justify an all-weather surface if it can be used for more than one purpose. There is no reason why a properly constructed all-weather surface cannot also be used as a surfaced holding yard (see the section **Surfaced holding yards** and Appendix: **The Equicentral System**). A surfaced holding yard can be quite expensive to construct, as can an all-weather riding/training surface, therefore it makes sense that, if it is possible, combining the two can work well for some people by saving them money.

If you do plan to use an all-weather surface as a surfaced holding yard, then a feeding and watering area can be constructed at one end or down one side. Large rubber mats can be used for feeding on so that hay and other organic matter does not get mixed in with the surface.

In addition, large hay feeders, hay nets, or both can be used. Smaller individual holding yards can be built around the all-weather surface for confining horses when the surface is being used to work a horse. These smaller surfaced holding yards will also be useful for feeding any concentrates, as horses do not share concentrates as amicably as they do hay.

If your budget allows it, a large roof can cover the smaller individual holding yards and the hay feeding area and can also extend partially over the riding/training area so that you have a partially covered all-weather surface!

There is no reason why a properly constructed all-weather surface cannot also be used as a surfaced holding yard.

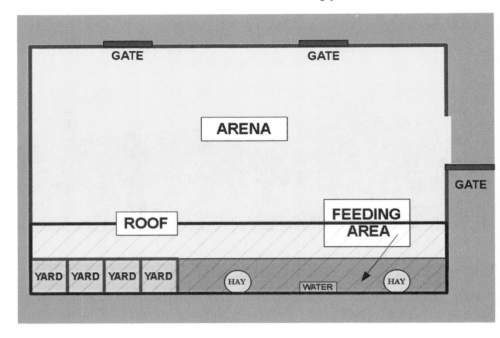

You will find that even though horses have access to the whole area, they spend most of their time loafing where the hay and water is, especially if they can see human activity from the same spot. Horses soon learn that human activity means they may be about to have their hay topped up, or better still, be fed concentrates, so they usually like to stand where they can see people. This is the main reason why horses that are being fed supplements spend hours standing at the gateway to a paddock, even when there is grass in the paddock. Therefore, if possible, make the 'feeding/loafing' area closest to the house or feed storage area.

You can also put a retractable electric fence partway across the arena if you feel you need to confine your horses to one end and this fence can then be retracted when you ride/train.

There is more about this system of management later in this book. See Appendix: *The Equicentral System*.

Other uses for an all-weather surface include using the area for occasional stock work with other animals such as cattle. You will probably need additional facilities such as a race, crush and ramps etc. but again, it is possible to double up and save expense.

Riding arena, training yard or both?

If your budget or the available space will accommodate an arena or a training yard, but not both, you need to think carefully about which would be best. A training yard usually costs less to build than a riding arena due to its smaller size. Keep in mind that you may not need as large an area as you think; you can do a lot with a surfaced 20m (66ft) square training yard (see the section *All-weather surface size and shape*). Conversely, if you do construct a larger area, you have the option of reducing its size to make a smaller area if required.

For example, even dressage riders can usually manage with a standard size dressage arena (20m x 40m or 66ft x 132ft) rather than an Olympic sized dressage arena (20m x 60m or 66ft x 198ft). It is only on the few occasions that you may need to practice a whole dressage test that the full size is needed. For those occasions, either a flat piece of land can be marked out, especially as your test at the competition may be on grass anyway, or you could travel your horse to a nearby Olympic sized dressage arena, which has the further advantage of getting your horse out and about and used to somewhere new.

If your budget or the available space will accommodate an arena or a training yard, but not both, you need to think carefully about which would be best.

Indoor or outdoor?

The climate in which you live also has some bearing on the answer to this question, although if you do not have the huge budget needed for an indoor arena (and that includes most of us), then this is probably a redundant question, even if where you live has extreme weather conditions. As well as protection from inclement weather, a roof protects against the sun in hot climates. Again though, unless the budget allows for such an expensive structure, you will have to time your outdoor pursuits so that you are not working with your horse when the sun is overhead.

This section of the book is mainly concerned with outdoor riding arenas and training yards; indoor surfaces have different base and surface requirements. For example, an indoor surface does not have to cope with rain so it can reflect this. Put another way, an outdoor all-weather surface will be fine if you put a roof on it later (as many people do), whereas if you were to take the roof off an indoor riding surface, it would probably not cope with rain because the construction of it would not have taken inclement weather into consideration.

If you think you may be going to put a roof on your all-weather surface in the future, you need to research roof truss sizes at this stage so that you will not have to make alterations to the size and shape of the surface later. If you do plan to cover your surface, keep in mind that a half or partly covered arena or training yard can work really well, is cheaper and gives you the best of both worlds. It means that you will have somewhere to retreat to in inclement weather and, on days that are really wet, you can simply stay in the area that is under the roof. However, even on wet days, it is no real hardship to ride out into the weather to some extent once mounted. This acclimatises your horse to coping with being ridden in bad weather in case it is ever necessary.

Sometimes a good compromise is to have a riding/training surface that is partially covered.

An indoor arena or training yard is of course out of the budget of many horse/property owners and this book is after all a *practical* guide to setting up and managing a horse property. That said, there are ways that it may be possible for you to end up with at least a certain amount of your riding/training surface under cover. Alternatively, it may be that even though you cannot afford to roof the area now, you may be able to in the future, so keep that in mind when planning this facility.

Sometimes a good compromise is to have a riding/training surface that is *partially* covered. For example, if you are planning on having a combined riding surface/surfaced holding yard as described in the section **Can this area be multi purpose?,** a large roof that provides shelter for the surfaced holding yard/s may also provide shelter for the all-weather surface (see Appendix: **The Equicentral System**).

Another cost saving option is to have a full roof on a training yard (rather than a larger riding arena).

All-weather surface size and shape

Riding arena size and shape

A riding arena can be any size or shape you like, but there are certain sizes and shapes that are commonly used for certain pursuits:

- A standard dressage arena is 20m x 40m (66ft x 132ft) and an Olympic sized dressage arena is 20m x 60m (66ft x 198ft). If space is at a premium, you may decide that the standard size is fine for training and you will hire a full sized one for practising the full test just before a competition. This option is dependent on there being one that you can hire and travel to in your area.

- For jumping, an arena will usually need to be at least 40m x 60m (132ft x 330ft) and may need to be as large as 60m x 100m (198ft x 330ft). It all depends on how seriously you follow your chosen pursuit (e.g. higher jumpers need a larger arena). Again, you may be better off creating a smaller every day working arena and just hiring a larger arena for certain activities.

- For driving, an arena of about 40m x 80m (132ft x 264ft) will be required.

- An all-purpose arena could be around 50m x 90m (164ft x 295ft), but can also be much smaller of course.
- Western pursuits require about 30m x 60m (100ft x 200ft) for reining and 40m x 60m (132ft x 200ft) for roping.

Some of these arena size figures are just a guide; unless you are already experienced in your particular chosen pursuit, it is important to speak with a professional (such as your coach) about what makes a good size and shape arena for your particular discipline. As well as the size and shape, the surface and fencing requirements change across different disciplines too, so find out as much as you can before making any decisions.

Training yard size and shape

Training yards can be round, square or a blend of the two. A 'round yard' is often used for 'starting' horses. For this purpose it is usually quite small e.g. about 11m (36ft) diameter. Unless you 'start' horses professionally, a training yard of this size is usually too small for general use, so this size would generally not be seen on anything other than a professional horse trainer's property. A more useful size for a training yard is between 18m to 22m (59ft to 72ft) diameter. As a guide, an 18 metre (59ft) circle is roughly the circumference required to lunge a horse.

Some people find a similar sized (about 20m (66ft) across) *square* training yard more useful because then there are straight edges for certain exercises, particularly close contact 'in hand' ground work. A square training yard can easily be made round if necessary by using panels or rails to temporarily round the corners off.

Yet another alternative is to have a training yard that is a *combination* of round and square with one or two right angle corners and the rest rounded off, or four straight sides with four rounded corners. Such a training yard can be very versatile.

Base and surface

Remember - this is only a guide to the subject. You still need to ask around and do more research. Also, as already mentioned, if you are planning on creating an all-weather surface yourself, you are better off starting with a smaller area, perhaps an area that can be used as a combined holding area and riding/training area.

An all-weather surface usually consists of a base and a top surface:

The base

The base is the foundations of an all-weather surface. Once an all-weather surface is completed, you can no longer see the base because it is covered by the top surface and therefore not seen, meaning that it is often unappreciated. The base is a very important parts of an all-weather surface – more important than the top surface in fact because if it is not constructed correctly, it makes no difference what the top surface is made up of, the project will fail sooner or later. The more time spent getting the base right, the better the finished all-weather surface will be. It cannot be stressed enough that an all-weather surface is only as good as its foundations.

This training yard, which consists of a permanent surface and lightweight portable steel panels, is a 20m (66ft) square with rounded corners! This can be a very useful size and shape.

Probably one of the most common mistakes made is economising on the base; something usually done due to inexperience, cutting corners or both. If the base fails, then not only does this have to be corrected, but due to mixing of the top surface and base, the top surface will usually be spoiled as well; a very common occurrence in failed all-weather surfaces. Economising on the base usually leads to a compromised all-weather surface that will require *much* more expense later to correct it.

It is easier to remedy if the *base is correct* but the top surface incorrect for whatever reason, as the top surface can usually either be added to (with a material that will improve it) or may, if necessary, be able to be removed without disturbing the base, allowing another, new surface to be applied.

The first stage of all-weather surface construction is earthworks. Earthworks should be carried out by an experienced professional and any topsoil should be removed and separated for use elsewhere. Topsoil should never be used as part of a 'cut and fill' for an all-weather surface, because topsoil contains organic matter which decomposes over time and reduces in volume. This decomposition may lead to subsidence and an uneven surface in the future. Any removed topsoil can be used to finish off the outside of your all-weather surface or used on other parts of your property.

The secret of a good base is thorough compaction and, in some areas/countries, drainage is also a feature.

The top surface should not go down until you are sure that the base is properly compacted.

Some all-weather surfaces are designed so that water runs into the top surface but then off the base. In this case, the base and top surface will have a 'fall'. The fall is usually from one corner to a diagonally opposite corner, or from the centre line to the outside edge on either side. About 1% from the centre line to each side should be sufficient in a normal rainfall area. If you live in a high rainfall area, it may need more (about 2%). On a large area where the slope goes from one side to the other, you may need 2% - 3% because the water has farther to travel. Drains (channels) at the sides of the all-weather surface can then take this water away if necessary. For an all-weather surface built into the side of a hill, you may decide to slope the whole area in the same direction as the hillside, meaning a little less excavation, and the water will then run off the all-weather surface and continue down the hill. A drainage ditch will still need to be constructed between the all-weather surface and the hillside.

An earthmoving contractor should have laser levellers that make sure the fall is correct and it may also be possible to hire a laser leveller from a hire shop. This fall should only just be enough so that water moves through and off the top surface, but not so much that the horses are working on a noticeable slope, or that water moves too quickly off the surface, because this will result in erosion of the surface.

This arena would have required extensive excavation due to the slope. In a case such as this a drainage ditch is required between the arena and the bottom of the slope as there will be water running off the hill.

Alternatively, some all-weather surfaces are designed so that water travels down through the top surface, into the base and into draining that is built into it. This type of all-weather surface may also have a slight fall but not usually as much as in the previous example.

An all-weather surface base can be likened to a road base. If a road base is not correct, then holes appear in the tarmac later. A good base starts with a coarse layer which is and thoroughly compacted. Finer layers are then progressively added and these are compacted. The base should extend at least 0.5m beyond where the outer perimeter of your all-weather surface will be to prevent crumbling of the outer edges when working on it.

A contractor will compact the surface over several days with very heavy machinery. If you are constructing the all-weather surface yourself, it is possible to hire a vibrating roller which you - and as many volunteers as you can rope in - can push around the base for several hours - or days. If you are building the all-weather surface in stages, time and rainfall will also help to compact the base. The surface should be compacted enough so that heavy trucks can drive over it without leaving any impressions. The hoof of a fast moving horse exerts far more pressure per square inch than a truck wheel, so if the surface won't take a truck, it will not stand up to repeated use by horses!

Constructing the all-weather surface over time means that any problem areas (soft spots, low spots etc.) can be noticed and fixed before the next layer goes down. Do not put the next layer on until you are sure the previous one is properly compacted. Good drainage is the key to preventing pooling of water (which leads to soft spots). The base will need to have a 'fall' so that rain water travels down through the top (riding/training) surface and then off the base. It is useful if there is a heavy rainfall between finishing the base and putting the top surface on. This way you can see where the water is going and check that the fall is correct.

Vegetation can be grown in an area that water runs through to slow it down, or even to divert it, before it gets to your all-weather surface. Likewise, any steep areas that are created due to the construction of the all-weather surface will need to be vegetated to prevent erosion. Plants such as grass, bushes and trees can be grown on the lower side or end of an all-weather surface to reduce the speed of any surface water. You want the water to leave the all-weather surface, but not too quickly, otherwise erosion will occur which will eventually undermine the all-weather surface.

Drainage on this professionally constructed arena is good, note the slight slope and drainage ditch down one side.

If your property is situated in a flat area, you could bring in materials that raise the base above the sounding land. If any road works are being carried out in your area, speak to the contractors. Old road base (large chunks of rock, tarmac etc.) makes an excellent all-weather surface base, and best of all it may be free! This material usually has to be dumped at a tip for a cost and the tip may also be farther away than your property, so a great arrangement may come to pass. The contractors may even be able do some work for you (e.g. levelling and compacting the base) while there, which will be cheaper due to them having their machinery in the area already.

A membrane may need to be used to stop clay or sub-soil moving upwards into the base. It is not always necessary to have one; they are not suitable for all surfaces and can rip and start to lift if incorrectly applied. You must also be very careful carrying out maintenance work on an all-weather surface that has a membrane, because anything that is dragged across the surface can catch on the membrane and pull it up, so make sure you get the right advice before using one.

Geo-textiles, large sheets made from a petrochemical based polymer product, and Geo-grids, plastic three dimensional grids, are often used in countries such as the USA and Canada to help keep the layers separate. Geo-textiles are placed between a sub-base and base and Geo-grids between the base and the top surface. They are not currently commonplace in countries such as Australia. Again, they may or may not be necessary and, because they will add a lot of expense to the final cost, you need to do your research thoroughly to find out whether they are necessary. You also need to research whether your chosen surface will work well with Geo-textiles and Geo-grids if you decide to use them, as some will and some will not - speak to the professionals.

The top surface

This is the part that you can see and that you ride/train on. **Materials used for top surfaces are varied and include:**

- **Wood products, woodchips, bark etc.** - there are many options for wood products as all-weather surfaces. They will decompose over time and can be slippery therefore this is not usually a good surface in wet climates. As they hold water they tend to freeze in cold climates. Hardwood chips will last longer than bark.

- **Sand** - should be washed (to remove soil) and screened (filtered to take out large lumps of rock etc.) before being laid. Sand can be dusty in summer and freeze in winter. Sand lasts a long time and is not usually as expensive as some other surfaces.

- **Rubber products** - long lasting but not cheap. Rubber may retain heat in hot climates (such as in certain parts of Australia and the USA) or may not be UV resistant. One problem with rubber is that it will break down over time and, as it breaks down, it will create finer particles, causing rubber dust which is sometimes harmful if breathed in. Rubber particles are also difficult to get rid of if you ever decide to change the surface altogether.

- **Commercially produced surfaces** - these can be a combination of all sorts of materials and may include wax, fibres, sand, sawdust etc. The mix and composition of these are very important. Poorly mixed products can lead to irregularities in your all-weather surface where components of the mix have clumped.

- Some products can be mixed e.g. rubber and sand, or applied in layers e.g. sand then rubber.

Sand is the most common arena surface. It is usually best to put down less rather than more, and top it up when necessary. Otherwise sand can be too deep and can be very hard for horses to move in.

Grass - the initial cost of this type of riding arena/training yard can still be considerable, depending on how much earthwork needs to be done to set it up, however once established, the area will be aesthetically pleasing and can add to - rather than detract from - your available grazing. Depending on the current soil type in the area, a base may still need to be constructed. Clay soil, for example, will not work well because it will tend to hold too much water. A well-draining sandy-loam may work depending on the percentage of clay content. Sandy soil will drain well and, if the grass is managed properly, the surface should be successful. If the area in question is already performing well when used for riding/training, it may just need to be improved so that it is smoother and so that it has the correct slope for drainage for example.

Commercially produced surfaces can be a combination of all sorts of materials and may include wax, fibres, sand, sawdust etc.

Well managed grassed surfaces have a low to no dust factor which is an important consideration in some localities. For example, in suburbia, dust can create a problem with neighbours. However, grass as a surface will negate being able to use the area as a surfaced holding yard as well so keep this in mind (See the section **Can this area be multi-purpose?**).

Well managed grassed surfaces have a low to no dust factor which is an important consideration in some localities.

Grass will often work when the surface is only used for several hours a week as opposed to several hours a day. Where most of the wear and tear occurs is also a factor. For example, dressage involves frequently riding the outside track (the area that is the very edge of the riding arena). This tendency is even higher in a riding school situation and is usually too much pressure for a grass surface to cope with. Show jumpers can also wear tracks leading to jumps etc. but periodic moving of the jumps can remedy this. Other equine pursuits tend to rely less on riding the outside track and are more likely to get away with a grass surface.

A hard or soft surface?

Whatever surface is used, it should not be too deep (soft) as this puts too much strain on the soft tissues of the legs of horses due to being very difficult to move in; think about how difficult it is to move in deep sand on the beach. Likewise, a surface that is too hard can cause concussion to horses if they are not conditioned to it; however it is generally best to start off with a reasonably firm surface. It is easier to add more surface than to remove too much surface. About 10cm (4ins) of a top surface is about the maximum you should need to apply, but 5cm (2ins) is often enough for a general purpose surface. If you are not sure, start with less and add more if necessary.

Factors that also determine the selection of the surface are your budget and the availability of materials in your area. **Other factors include:**

- The climate e.g. how much rain, sunlight/heat and wind the surface will have to contend with. High prevailing winds will bias surface selection toward the heavier materials, while high levels of sunlight and heat will bias selection away from rubber composite surfaces that are more prone to UV degradation.

- The availability of water. Dust is often a major problem with all-weather surfaces and can understandably alienate otherwise compliant neighbours. Choose a surface that produces as little dust as possible. Using lots of water to dampen an all-weather surface is not sustainable either, so aim to have a low dust surface to start with. Pool salt can be used as a dust suppressant as it absorbs and holds moisture, therefore reducing the amount of dust. It has the added advantage of also being a weed suppressant, so sprinkling pool salt in areas prone to grass or weed growth, such as the corners will kill out most weeds. It is also possible to buy commercially produced dust suppressants.

- Another consideration is that, in time, you may have to remove the surface due to poor performance, change of use and because surfaces also degrade and wear out. A surface that can be spread on land after use (e.g. sand, wood products) rather than have to be taken to a tip (e.g. rubber) is preferable.

A surface should be firm and not too deep, although different disciplines have different requirements.

Edging the top surface

Depending on how your all-weather surface is constructed, you may need edging around it to keep it from spilling out into the surrounding area. Edging can consist of railway sleepers, pine poles etc. however, turfing the surrounding area to the level of the top of the surface may be just as effective, more aesthetically pleasing and safer. Use a non-creeping variety of grass unless the all-weather surface is in heavy use, otherwise it will constantly invade the all-weather surface and will need to be removed continually. Any topsoil that was removed during the earthworks can be spread around the outside of the all-weather surface creating a neat finish. This area can be seeded with grass. A low solid barrier is better if used in conjunction with a higher fence because falling riders can easily be injured when landing across something that is low to the ground. Remember, any edging must not trap surface water; the all-weather surface still needs to drain.

Fencing your all-weather surface

A riding arena can either be fenced, or left unfenced. Whether you choose to fence it or not depends on what you use if for, but if you are teaching beginner riders/children then the area should be fenced. You would, of course, fence the area if you are also using it as a surfaced holding yard. Training yards are usually fenced so that horses can be worked loose or ridden.

If you plan to also use your all-weather surface as a surfaced holding yard (see the section *Can this area be multi-purpose?*) or even just for occasional turnout for horses, the fence must be high enough and strong enough to keep them in. Since an all-weather surface is a small area compared to a paddock and therefore horses are more likely to challenge the fence, consider a minimum height of 1.4m (4.6ft) for a 15hh horse. Make sure you also read and cross-check the information in the sections *Surfaced holding yard fences* and *Fences and gates on a horse property.*

An all-weather surface that is used for teaching beginner riders and children should *always* be securely fenced. The fence needs to be solid, e.g. post and rails or similar, so that if a horse gets out of control, or a rider falls off, the horse is still confined in the area.

All-weather surface fence height

The required height of the fence depends on what the all-weather surface is used for; if an all-weather surface is to be used for ordinary riding only (no serious jumping), then a fence of 1.2m to 1.4m (3.9ft to 4.6ft) should be fine. An all-weather surface that is to be used for jumping should either be left unfenced, or fenced higher than the height of the jumps. Alternatively, the jumps should be placed well away from the fence. A professional show jumper usually has a very large all-weather surface or a dedicated paddock for jumping large jumps, but for most average riders, a 1.4m (4.6ft) fence is adequate.

A smaller area (e.g. a training yard rather than a riding arena) usually has a higher fence. A good *minimum* height to aim for is that the top, or baulk rail, should be about waist level with a mounted rider (about 1.8m to 2m/5.9ft to 6.6ft). Make sure the tops of the posts are not higher than the top rail, as projections can be dangerous. Rails should be on the inside to prevent riders (and horses) from banging into protruding posts, or the inside should have a smooth surface e.g. rubber or smooth ply panels fitted to the inside. The posts can be set at a distance of 1.8m to 2m (5.9ft to 6.6ft) apart however it does not really matter unless it affects the fence strength.

A smaller area (e.g. a training yard rather than a riding arena) usually has a higher fence.

All-weather surface fence materials

Any fencing needs to be secure, highly visible, low maintenance (preferably) and, most importantly, safe for both riders and horses. This rules-out steel ('star picket') posts, due to the risk of impalement, electric fencing of any kind, (for obvious reasons!) and any other material that could injure a rider or a horse. Another big 'no-no' is chain or rope, because if a panicking horse runs through it, the whole fence can become tangled in their legs and dragged with them as they go. You should also avoid ornamental fixtures such as concrete plant holders that can cause serious injury to a falling rider or horse. That said, a solid post and rail fence, while deemed safer, can also cause injury in certain circumstances.

Keep in mind that materials such as rubber to ground level will help to keep the surface in but it also tend to keep water in when it rains.

The options for fencing include post and rail, hedging/natural vegetation, steel pipe, portable steel panels, horse mesh etc., with the main consideration being safety for riders and horses **therefore you need to think about:**

- How safe the will be fence if a horse were to run in to it.
- How safe will the fence be if a horse were to try and jump it.
- How visible the fence will be.

The advantages and disadvantages of different fence types are discussed in detail in the section *Fences and gates*. The following section contains some additional information about certain fence types when they are being used for an all-weather surface in particular, because in that situation, riders and horses are coming into close contact with the fence, meaning that there are some different issues to consider than is the case when a fence is simply used to keep horses in a paddock

Hedging/natural vegetation - advantages:

- It is usually the best medium to hit at speed! However, you should avoid trees that will mature with large trunks - tall dense bushes are better.
- It is a solid looking barrier and, once it has grown tall enough, even an out o control horse will not usually attempt to go over or through it.
- It is inexpensive however it takes time to grow.
- It provides a windbreak for the riding/training area.

Hedging/natural vegetation - disadvantages:

- It may prevent the top surface from drying out quickly enough after wet weather by preventing the sun and wind from getting to the top surface.
- It may prevent riders/handlers from being observed from other parts of the property - a potential safety issue – ideally people working horses should always be able to be observed by others in case an accident occurs.

If the all-weather surface is to double up as a surfaced holding yard, you will need to do some research about what plants would be best to use (and that suit you locality). You need something that the horses will either ignore or that they can browse safely, but not annihilate in one sitting! A more solid fence may still be required between the vegetation and the all-weather surface particularly if you are also using the area as a surfaced holding yard, but the vegetation will make the solid fence much safer and a much better visual barrier for a horse travelling a speed. This vegetation has the added benefit of providing habitat for wildlife, some of which will eat flies and mosquitoes.

A solid type of fence backed up by vegetation can be an excellent fencing choice for an arena.

Portable steel panels

These are (usually) steel panels that fasten together and may also have rubber sheets attached to them to make the inside surface solid and smooth. The panels can be unfastened and moved if necessary.

Portable panels have several advantages over fixed fencing for an all-weather surface:

- They allow the shape and size of the area to be adjusted.
- They can be taken with you if you move to another property.
- They cost about the same as a solid fence.

Portable panels range from very lightweight, lightweight steel and even plastic panels are now available, through to very heavy, usually a combination of small aperture mesh and marine ply in a steel frame. They are usually around 1.8m to 2m/5.9ft to 6.6ft high, making them ideal for smaller areas, e.g. a training yard rather than an arena. There is no reason why they cannot be used for a larger area, but the more solid types of portable panels may work out very expensive in this case.

Steel panels are becoming increasingly popular due to their versatility.

Keep in mind that certain commercially made steel fence panels such as those made for cattle have rails that are too close together for horses, so if you are also using this area as a surfaced holding yard, keep the following information in mind. A horse will twist their head and neck to put their head through a narrow gap, but can forget to do this when pulling their head back – especially if they are rushing or panicking. This can result in serious injury. Aim to prevent horses from ever putting their head through a small gap of any kind. Small aperture mesh can be used to prevent a horse from pushing their head through gaps.

A novel and inexpensive option for an all-weather surface is to use wooden posts with plastic 'polypipe' rails.

A round training yard can be fenced so that the wall/fence leans out. This is to reduce the chance of a mounted rider banging their legs. However, the disadvantage of this style of fence is that a horse's hooves will knock the sides more easily, because their body – the widest part of them – is further to the outside of the circle. This can actually cause injuries and can even cause a horse to fall. This style of fence is also more difficult to build and to maintain.

Training yard - solid walls/fences or open?

Some horse trainers prefer 'solid walled' training yards and some prefer open fenced training yards A solid wall prevents a horse from being distracted by what is going on outside the yard, however solid walled training yards can be dangerous if a person needs to get out quickly. They cost more to build and also concentrate heat in the area, reduce airflow, and reduce the amount of sunlight getting in to the area, which would otherwise help to dry the surface after wet weather.

This yard is fenced inexpensively but effectively with wood posts and plastic pipe. Notice the slight lean.

Open style post and rail or pipe/panel training yards are cheaper to build and maintain. The bottom rail should be no less than 30cm (1ft) from the ground to reduce the risk of leg injuries to fast moving horses. A rubber strip can also be fitted around the bottom edge to make it safer.

Again, a solid 'lip' around the bottom of a training yard will help to keep the surface in, but will also prevent water from leaving the surface easily after rain. You can use old railway ties (sleepers) at the base of your training yard for this purpose. Old railway ties can vary in length slightly, so make sure you space your posts accordingly (before putting them in).

A cheaper way of keeping the surface in place is to grow dense vegetation (such as lawn type grasses) around the outside of the surface. The ground around the outside of the yard needs to slope slightly away so that the training yard base is the highest point. This will allow water to leave the surface immediately after arriving, and the vegetation will filter it and keep the surface material in.

All weather surface gateways

A gateway on a fenced all-weather surface should be able to be opened by a mounted rider. Therefore, the latch should be operable from both sides and the gate should ideally swing both ways. It should be flush with the fence and there should be no gaps between the post and the gate when shut. It must also have no projections (including the gate latch) and the gate needs to be wide enough to get machinery into the area for maintenance (therefore not less than 3m/10ft). Two separate gates can be installed, a wider one for machinery and a narrower one (but not less than 1.2m/3.9ft) for people/horses.

Some horse trainers prefer 'solid walled' training yards and some prefer open fenced training yards.

A gateway on a fenced all-weather surface should be able to be opened by a mounted rider.

All-weather surface lights

Being able to ride/train after dark may be necessary for you. Professionally erected lights can cost several thousands, in fact they can cost almost as much as the all-weather surface itself, therefore get as many quotes as possible if this is the way you plan to go. Ask to speak to their clients if possible and find out if they are happy. You may be able to do some of the work yourself. You will of course need an electrician for the electrics, but you can dig trenches (with a hired trench digger) and lay conduit yourself. You can also hire a post-hole digger to put the posts in. If you do an Internet search on this subject, you will find examples of people who have erected lights themselves with examples of costings etc.

Make sure that outdoor lights are not directed towards neighbouring properties, in particular their living spaces.

Having lights means that you can work your horses after dark. They can be very expensive so if you do not have a large budget but you need them, you may have to DIY some of the work.

All-weather surface maintenance

All-weather surfaces are meant to be just that; you should be able to ride/train on them in any weather within reason. Some all-weather surfaces require a lot of maintenance e.g. daily watering to keep the dust down. Obviously, it is better if your all-weather surface is low maintenance. However, if you do not take care of it, it will degrade over time.

Some does and don'ts for maintaining an all-weather surface are:

- **Don't** allow weeds to invade the top surface. Weeds may be growing up from the base, but this is unlikely if the base was compacted properly. More often, weed seeds arrive on the wind or via bird droppings. They also may be in manure piles left on the surface. Weeds quickly spread and, even if you kill the plants, they will leave organic matter underground, which will change the properties of the base and top surface. If your all-weather surface has a membrane, pulling at established deep rooted weeds can pull up this membrane, so keep on top of weeds and remove them before they become established. A sprinkling of pool salt can help to suppress weeds. One advantage of using your surface as a holding yard is that it will not become overgrown with vegetation.

- **Don't** allow low areas which will lead to pools of water sitting on the top surface to develop, as these pools will create boggy areas and eventually lead to surface failure. Either shovel some spare top surface material from a corner or edge - where top surface material tends to build up - or use a proper all-weather surface maintenance implement to smooth the top surface again.

- **Don't** use a pasture harrow or any other implement which is designed for pasture on your all-weather surface (unless your all-weather surface is grassed) or you will damage it.

- **Don't** ride/train on the all-weather surface during or immediately after heavy rain; wait for the main body of water to run off. Even though it is an all-weather surface, you still need to give it time to recover.

- **Don't** leave manure or any other organic matter such as leaves and dead weeds on the top surface – pick it up daily, especially if dung beetles are working on your property. Dung beetles take manure underground rapidly. If you leave a pile of manure today - it will be underground tomorrow. Dung beetles are great for your land, but not for your all-weather surface (unless the surface is grassed) (see *The Equicentral System Series Book 2 - Healthy Land, Healthy Pasture, Healthy Horses* for detailed information about manure management including information about Dung beetles).

- **Do** regularly rake or shovel the edges and corners back into the centre using a product specifically designed for all-weather surface maintenance. You can buy an implement that can be pulled behind a vehicle and levels the top surface. Aim to carry out this levelling at least once a week. However, in a high use situation it may even be necessary to do this daily.

- **Do** top up the top surface as soon as it becomes necessary, otherwise you will be riding on a layer that is not meant to be ridden on, leading to more problems in the future.

See the section **Horse facility positioning** for information about the *positioning* of an all-weather surface.

—

You can buy an implement that can be pulled behind a vehicle and levels the top surface.

Chapter 4: Horse facility planning

Design plays a very important role in our lives; good design leads to better living and working spaces. It is therefore very important that we look at our horse property as a whole, with a view to creating a design that will work for our chosen lifestyle, our chosen horse pursuit/s, keep our horses healthy and happy, enhance the environment, and to be pleasing to the eye, all at the same time.

There are many benefits to spending time on the planning stage so that you end up with a good horse property plan:

- The correct positioning of fences, laneways, buildings, surfaced holding yards and other horse facilities is essential for the successful operation and management of a horse property.

- If it is well planned, the property will be a safer, more productive, more enjoyable place to be with horses.

- At the same time it will be labour saving and cost effective due to better efficiency.

- It will be more aesthetically pleasing and therefore a more valuable piece of real estate.

- If the property is also a commercial enterprise, a well-planned property will be a boon to your business.

- A well-planned property is also good for the environment.

Designing your own horse property can be very rewarding, but it can also be very frustrating because if you make the wrong decisions at the planning and implementation stages, you will have to live with these poor decisions, spend more time and money making alterations, or sell the property and start again! More time spent in planning and preparation leads to more informed decision making so take your time.

This book is a guide only -make sure you get as much advice as possible from as many sources as possible before committing to any expensive projects.

When planning a property, remember it may take many years to achieve your goals. Enjoy the journey, but be flexible and accept that your own ideas may change as you gain experience.

Making a plan

You are now at the stage where you can start to plan the property in earnest. After reading this book and speaking with other people, you should now have many ideas about what you would like to do with the property. Whether you are designing a property from scratch or improving an existing property - planning is a crucial step in the process. It is too easy to rush off and plant trees and have fences erected, only to realise that you have put things in the wrong place; it is better to minimise wasted time and money by doing things right the first time.

Your plan should be accurate and detailed. The best starting point for a good property plan is an aerial photograph and, these days, these are readily available through local authorities, private companies and even Google Earth, although not always up to date or detailed enough in some areas.

By using this photograph and tracing the shape of the land block and its current features on to a sheet of paper or a transparency, a series of plans can be made at various stages of property development. If you are unable to obtain an aerial photo, then sketch the shape of the block onto a sheet of paper.

The property features that you need to mark on this plan include:

- Landscape features such as valleys, slopes, gullies, ridges, contour lines, trees and shelterbelts and water (farm dams, streams/streams, and rivers).

- The direction of prevailing winds and the angle of the sun in summer and winter where the sun sets and rises etc.

- Anything on the outside of the property boundaries that could affect the property such as neighbours (and their proximity) a dirt road (dusty and noisy), an intensive farming operation (possible chemicals, smells from sheds etc.).

- Different land classes, for example rocky, swampy, heavy clay and sandy areas (even small properties usually have different types of land, especially if the property is undulating).

The best starting point for a good property plan is an aerial photograph.

- Problem areas such as erosion (tunnel, water or wind erosion) waterlogged soil, compacted soil (areas that animals tend to congregate), habitat for pest animals (e.g. rabbit warrens) and weeds.

142

- Fixtures such as buildings, existing horse facilities, power lines, easements, wells, fences and drainage lines.

- Any zoning requirements that will affect where you can or cannot build, for example part of the property may be classed as water catchment, other areas may contain flora or fauna of ecological significance.

Make up plenty of copies of this basic plan so that you can try out lots of different designs. When developing the property plan, seek all the advice available and, once you have your completed plan, aim to show it to people who have some experience of owning and designing a horse property and get their opinion, although keep in mind that their ideas may be traditional based and not necessarily ideal in terms of good environmental management. It is also a good idea to visit as many horse properties as possible to get ideas.

This plan (below left) shows a sloping property's existing features, including large bare and compacted areas and erosion created by inappropriate past management. Notice that these bare areas are where the horses spend a lot of time loafing where they can also see the facilities or house.

This plan (below right) shows a better layout for the same sloping property. This layout can utilise **The Equicentral System** *(see Appendix:* **The Equicentral System***) or water troughs can be put in each paddock for a more traditional management system. In either case the header tank will gravity feed the water trough/s.*

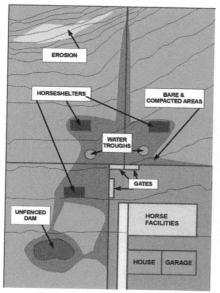 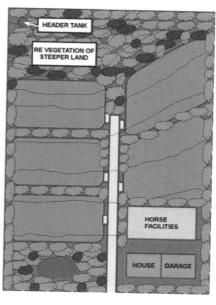

Building permits

Permits are usually required for any significant sized buildings on a horse property In some localities, the erection of simple horse surfaced holding yards and shelters, or the construction of an all-weather surface or farm dam will also need permit. What is regarded as a significant building or as significant earthwork varies from area to area, so check the current regulations and requirements for permits in your own area before starting construction, because you may be required to remove whatever you have constructed if you do not get permission first

In order to get a permit for any development work, you may need to fill in a application form and submit a site plan along with any engineer's computations builder's specifications (structural details) and plans showing the floor plan elevations, footings, etc. if applicable. If you are using a professional to erect the building/s,they may do this for you as part of the service. Bear in mind, however that you may be required to explain how the project will affect the water flow on the land and other environmental issues.

Acquiring the proper permits is just one part of a successful project. Take time at this stage to get things right so that you do not encounter problems later on due to incorrect preparation.

Permits are usually required for any significant sized buildings on a horse property.

Options for construction

Whatever facilities you decide to construct, there are many options open to you with regards construction:

- You can have the facilities built by professionals, therefore saving you time and labour – but this will be the most expensive option.

- You may be able to build the facilities yourself, which will be labour intensive and take time but can be very rewarding and cost effective *if* you have the expertise and equipment.

- A third option is to have a contractor carry out some of the work (the parts that you would find most difficult), and then do some of the finishing off work yourself.

You may be able to build the facilities yourself, which will be labour intensive and take time but can be very rewarding and cost effective if you have the expertise and equipment.

Buildings

There are many options available to you; you just have to decide what suits your particular situation. Whatever you choose to do, you need to have a good understanding of what constitutes good horse facilities.

Options for buildings include:

- Buying a shed/barn in 'kit' form and erecting it yourself.

- You could have a shed/barn erected by a professional and then build enclosures inside the building yourself. For example, you could construct the interior walls and partitions yourself, but it is also possible to buy prefabricated partitions that will turn a shed/barn into individual stables/surfaced holding yards etc.

- You could contract a builder. If using a general builder, aim to find someone that has been recommended by someone who has used them to build horse facilities and/or use a commercial horse facility building company.

Remember that the cheapest quote is not necessarily the best. In fact, in the case of horses, it is rarely the best. You need to check how many fixings they plan to use and the quality of the fixtures and fittings (which *must* be heavy duty otherwise they will not last and may even be dangerous).

Builders that do not have experience of building horse facilities (and of horses in general) tend to hugely underestimate the strength of horses and the amount of wear and tear they give. They also do not realise how easy it is for horses to injure themselves, therefore they may not realise the importance of having smooth edges and no tight spots that horses can trap parts of themselves in.

Fences and gates

The cost of fencing varies greatly so it is best to shop around before buying. **There are several alternatives open to you:**

- You could buy the materials and do the fencing yourself; this will be the cheapest option, but will only work if you have the time, inclination and skills. Fencing can be a very rewarding project to take on, but if you have no previous experience, try to find a short college course that will teach you some skills before attempting to fence a horse property.

- You could get a local fencing contractor to put the posts in for you and do the rest yourself. This is a good option, as the heavy work is done for you, which is very difficult without the correct machinery.

- You could get a local professional fencing contractor to erect them. A fencer may also have better buying power when it comes to obtaining fencing materials so compare prices and decide whether it is better if you supply the materials or they do.

- You could get a larger commercial fencing company to erect them. For example one of the companies that promote, supply and erect a certain type of horse fence (such as PVC fencing or 'horse mesh' fencing). This will probably be the most expensive option, however that way you will be able to get a guarantee for both the workmanship and the materials from the same place.

If you are erecting the fences yourself, try to buy in bulk so that you can negotiate a better deal when buying materials.

If you are erecting the fences yourself, try to buy in bulk so that you can negotiate a better deal when buying materials.

Likewise, if using a contractor, it is relatively more cost effective to get larger amounts of fencing done rather than small amounts. So, try to plan your budget so that you can afford to have significant amounts of fencing done at one time. When costing fencing, you need to work out what the fence will cost per metre when erected so that you can compare different options with each other. Don't forget to include all fixings as these make a big difference to the final price. You also need to take into account any on-going maintenance further down the track.

All-weather surfaces

Constructing an all-weather surface will be an expensive exercise whether you use a contractor or DIY. The total cost will vary, depending on factors such as the amount of earthworks required, the size of the all-weather surface and the type of top surface. The materials used for the base and top surface of the all-weather surface make up a large percentage of the total cost (along with the earthworks). You may find that, after costing the materials – including hire of any machinery – you are better off paying a professional to do the whole project. Another factor to consider is that a contractor may be able to get the materials at a better rate than you will because they are 'in the trade'.

If you decide to have the all-weather surface constructed for you, it is generally better to use a person or company that specialises in *equine* all-weather surface construction as opposed to a person or company that simply does earthworks for example. This is because specialist knowledge is required to construct surfaces that are suitable for horses to work on.

If you do decide to use someone who is not experienced, you need to guide them every step of the way. It is not uncommon for an inexperienced earthworks contractor to simply to clear and level an area of land and put sand on it, thinking that this will work. By the time a property owner realises that this surface has failed, it is already too late and the sand will have mixed with the uncompacted soil (which may contain clay) underneath. This ruins the sand surface, meaning it will have to be removed. You can see how it can cost a lot more if you do not get it right the first time.

Shop around and obtain several quotes before choosing a contractor and ensure you get any quotes in writing. Beware of any very cheap quotes, as this may mean that the contractor is inexperienced and does not know what materials and work are required to properly construct an *equine* all-weather surface, and that it is therefore likely to fail in the future. As with most things in life, you tend to get what you pay for, and companies that specialise in building all-weather surfaces should provide a guarantee on their works. Ask to see examples of all-weather surfaces they have already constructed and ask if you can speak to past clients.

Is it possible to build a good all-weather surface yourself even if you are not experienced? By doing as much as you can yourself, it is possible to save some money and learn a lot in the process. However, just remember that if it fails, you will have no comeback. If you are planning on doing this, it is a good idea to start with a smaller training yard rather than tackle a large arena in order to gain experience.

One option is to use a contractor/s for some of the work, particularly the earthmoving and levelling. You will need to do this anyway of course unless you already own earthmoving equipment.

A factor to keep in mind is that the earthworks required for all-weather surfaces can be extensive (and expensive) depending on the site e.g. the amount of slope and whether there is rock involved etc. Try to budget so that you can have all of the earthworks done at the same time if possible. This is because the cost of moving the large machinery to your property is usually a large percentage of the total earthworks cost (unless your neighbours have an earthmoving business!). So, while this machinery is on your property you need to make the best use of it.

If you are inexperienced, building an arena, or in this case an indoor roundyard, may be best left to the professionals. However with careful planning and preparation it is possible to DIY. Another alternative is that you do some of the work and a professional does the rest.

An earthwork contractor will be able to do other earthworks while at your property, for example while levelling an area for a building, so one option is to get them to level an area for an all-weather surface even if it will be a while before you can afford to finish it.

If water is a problem in your area, you will need to time the construction so that large vehicles will arrive during the dry time of year, or you may end up with trucks bogged on your land. Make sure that trucks will be able to gain access to the area, even after completion, so that the surface can be topped up in the future.

A professional all-weather surface contractor can usually construct a riding arena or training yard in about a week. However, there may be a long waiting list. Some may not be able to start on your project for several months or longer due to being in high demand. An advantage of building the arena or training yard yourself is that you can do the work in stages, as funds become available.

You can save money if you DIY, however it may be more expensive in the long term if the area has to be repaired later due to your inexperience. An all-weather surface built by a professional company can also fail, but at least in this case you should have a guarantee to fall back on.

A surfaced area for riding and training is a large expense if built properly - and even more so if done incorrectly - because even more time and expense will be needed to fix it. So, it is important that you get it right the first time. Good construction is equally important however much you use the surface.

The information in this book will help you whether you are planning on constructing the surface yourself or having a professional to do it for you, because the more you know, the better decisions you will be able to make. If you are planning on constructing the all-weather surface yourself, you will need to do quite a bit more research. This book is just a guide to this subject so that you have enough knowledge to make decisions about what your next step should be.

Constructing an all-weather surface will be an expensive exercise whether you use a contractor or DIY.

The planning framework

When planning a horse property, there are many things to think about:

- How many horses does the property need to support?
- What do you plan to do with the horses?
- What horse facilities will you need?
- Where will you put them?
- How many paddocks will you need?
- Where should the fences go?
- What is the best order for development?

It is easy to become overwhelmed by the sheer amount of decisions that you have to make; therefore it helps to have a framework for those decisions. The following seven factors will help you with this. They are sometimes interrelated but, by separating them into roughly seven groups, you should be able to make sure you have all of the important points covered.

How many horses does the property need to support?

The environmental factors

Achieving healthy environmental management of the land on which your horses live is as important as caring for your horse/s. This requires good planning, but will lead to a better 'lifestyle' for your horses and also boost their wellbeing. Horses thrive best in an environment that is as ecologically diverse and as close to 'natural' as possible. This means that horses should have access to pasture for at least part of each day, clean water, the companionship of other horses and shelter.

How can your horse property work with the environment rather than against it?

By taking care of these issues, you will also be enhancing and caring for the environment at the same time. Horse care and wellbeing goes hand in hand with good environmental management. **For example:**

- Pasture, at the same time as feeding your horses, provides habitat for other animals. The plants protect the soil from erosion and healthy soil in turn helps to grow more pasture. A variety of pasture plants, which is what horses need and prefer, helps to create a healthy ecosystem by providing habitat for all sorts of beneficial creatures. By sustaining vegetation cover through good planning and grazing management e.g. planning to have areas that horses can spend time in order to reduce grazing pressure on the land, you will avoid the problems of mud and dust. These conditions cause acute and chronic health issues such as respiratory disorders (dust), eye problems (dust) and skin abnormalities such as greasy heel (mud) and slipping injuries (mud).

- Clean water is obviously a very important requirement for your horses, but at the same time it also provides habitat for other beneficial creatures, as does the vegetation that immediately surrounds a waterway. This vegetation in turn helps to keep the waterway clean by filtering out nutrients that would otherwise end up flowing into it.

Clean water is obviously a very important requirement for your horses, but at the same time it also provides habitat for other beneficial creatures.

- By keeping horses in herds rather than separately, you will be able to achieve better land management, because separated horses lead to land management issues, such as the erosion caused by 'fence walking'. Horses can become fence walkers when they are separated, because they are not meant to live alone and tend to become stressed when they are forced to do so (more on this later).

- Vegetation grown for shelter also provides habitat for wildlife and some of these species will help to control pest insects for example. Vegetation also helps to control the 'micro climate' by cooling an area when the weather is hot and slowing down wind in inclement weather.

These are just some of the many benefits that come about when you take care of the environment, so it is important that you plan how the environment will affect the property and vice versa. Planning to have the least possible negative impact on the environment and, where necessary, enhancing the environment, should be a major concern. **So think about:**

- How your property can work *with* the environment rather than against it.

- What the best way to manage the flora and fauna on the property is.

- Where the best place to plant vegetation for shade is.

- How you can take care of the waterways.

These points, along with many others, need to be addressed when planning your property. These situations and many others like them, when correctly managed, become a win – win for people, horses and the environment alike. Factors that help the good environmental management of a property are included in every section of this book.

The horse welfare factors

When planning a horse property, it is easy to get carried away and plan to build facilities that suit yourself, but do not necessarily suit a horse's needs. How do we know what horses need? We have to consider how they live in their natural situation and go from there. We know that domestic horses are capable of surviving as natural living horses. Ever since their initial domestication, horses that have been released or have escaped into the wild have generally survived and thrived. This tells us that all of their natural survival instincts and behaviours are still intact and it therefore makes sense to work with these natural behaviours rather than against them.

We also know that when animals are prevented from carrying out their natural behaviours, they become stressed. Zoo animals and domestic animals such as horses, dogs and cats can exhibit physiological and behavioural problems if their 'lifestyle' is not right. In horses the *physiological* problems that tend to occur take the form of gastrointestinal problems such as colic and gastric ulcers. These can come about from a lack of adequate fibre. The *behavioural* problems (stereotypic behaviours) that they exhibit include 'crib-biting'/'windsucking', also from a lack of adequate fibre, and weaving, from a lack of movement and stimulation.

In the horse world, stereotypic behaviours are inaccurately termed 'vices'. The usual response to these 'vices' is to put a horse that displays them under even more stress by trying to *prevent* the horse from carrying out these behaviours, often involving the use of windsucking collars and anti-weaving bars for example. The assumption is that the horse is misbehaving and needs to be restrained or punished. Instead it should be understood that the horse in question is under stress and is reacting to that stress.

A common horse facility that can cause stress if designed and used incorrectly is a stable. Many people assume that they and their horses *need* stables, when in fact stables are often built more for human convenience than for the mental and physical wellbeing of horses. Stables *can* have uses on a horse property, but they are not essential.

—
See *The Equicentral System Series Book 1 – Horse Ownership Responsible Sustainable Ethical* for a discussion about how and why stables evolved. See also the section *Why stables?* in this book.
—

Previously domestic horses are very capable of thriving and surviving in the 'wild'.

In reality, horses are healthier if they are outside as much as possible. Horses actually thrive better in cold weather than in hot weather. If a horse cannot live at pasture full time for whatever reason, for example not enough grass/too much grass, land too wet/land too dry, then the next best thing for both the mental and physical wellbeing of a horse is an outdoor and fenced enclosure, a surfaced holding yard, that has a surface and shelter.

Horses need to live in a stimulating environment in order to thrive and should have daily turnout into either a grassy paddock or large areas where they can be with other horses for interactive social behaviour such as mutual grooming and play. They should also be given large amounts (preferably 'ad-lib') of *low energy fibrous feed* e.g. hay as opposed to small amounts of high energy feed. Good planning and *good horse management* should ensure that stables or surfaced holding yards do not become dungeons for horses. By using them appropriately, stables can have their uses, as can surfaced holding yards with shelters, so make sure you think about the pros and cons of each option before deciding what to build.

Aim to keep these facts in mind when designing your property so that you do not end up building expensive unnecessary facilities that, while fitting the traditional picture of what a horse property should look like, may not be adequate in terms of best horse welfare practices.

Our view of what a horse property should look like...

...is often very different to what a horse's view of a horse property should look like!

Your budget

Your budget dictates what can you afford to do and when you can afford to do it. Throughout this book, it is assumed that you have a strict budget to adhere to as the majority of people do not have bottomless pockets.

Planning will help you to prioritise projects; most people have a budget and, in many cases, just buying the property in the first place uses up all available funds for the time being. You need to separate your long term vision (what you want) from what you, your horses, your property and the environment, needs.

So, what should you plan to build first? For your own sake, after the house, somewhere safe to work and handle your horses will be a priority. For the sake of your horses, water, food, companionship and shelter are essential for their wellbeing. For the sake of your property and the environment, protection from land degradation and care/protection of natural features are a priority.

Many people start with the fences, boundary and internal, followed by the construction of a riding arena and stables. The pasture and other vegetation often gets left until much later and, in many cases, the property owners run out of steam and money, before getting around to these very important areas of the property.

Your budget will help you to separate what you would like from what you can actually afford to build.

A more efficient approach is to prioritise the *boundary* fences (the perimeter fences) as these are very important, but leave the *internal* fences until later. Boundary fences keep your animals safely on the property, and allow you to sleep well at night, knowing that your animals are not going to get out on the roads. The next priority depends on whether the property is being developed as a commercial property or not.

With a *private use* property, the next priority should be constructing an area that the horses can be confined to for shorter or longer periods. This could be surfaced holding yards with shelters or you could simply decide to create a *short term* 'sacrifice area' (see **The Equicentral System Series Book 2 - Healthy Land, Healthy Pasture, Healthy Horses** for information about sacrifice areas).

Internal fences can be erected using temporary electric fencing, as this is both economical and effective for horses. Temporary electric fencing makes pasture improvement easier as it can be moved or removed to allow any contractors with machinery to do their work. Indeed, it may even be best to always have movable internal fences for the flexibility they provide (see the section **Fence and gateway positioning**). This measure will free up money that would otherwise be spent on *permanent* internal fences.

The boundary fences are the most important priority on a horse property. Internal fences can come later.

The next priority should be improving and increasing vegetation on the property. These improvements will actually start to save you money quite quickly, which will enable you to get the things that you want (such as an all-weather surface) sooner. Paddocks with good grass save a huge amount of money otherwise spent on feed. Planting trees and bushes on the property for wildlife habitat and the many other benefits that they provide, should also be done as soon as possible. You, your animals, the local wildlife and the environment will then reap the benefits sooner (see the section **Vegetation planning**).

A *commercial* property usually needs to earn an income ASAP, so whatever facilities are needed for this purpose usually have to come immediately after the boundary fences. Therefore, a riding surface and horse holding facilities may be the next highest priority. Once these are in place and the business is starting to earn an income, further improvements to pasture and other vegetation on the property should follow soon after. Again, try not to rush into constructing permanent internal fences for the same reasons as outlined above. It is possible to achieve a neat and professional finish using electric fencing. Another alternative is to construct permanent fencing in areas that will be seen and used by clients and use temporary electric fences elsewhere.

As soon as the pastures are established they can start feeding the horses and you can start saving money for other things.

Keep in mind that many horse facilities such as round yards, stables, shelters etc. are now available as portable items so that you can try out different configurations before deciding on their permanent positions.

The ergonomic factors

Utilising ergonomics is about organising a working environment for ease of use and safety. Ultimately, good ergonomics leads to a better lifestyle as good design reduces hard work, accidents and stress. Good ergonomics are essential for a private use as well as a commercial property. You need to think about what will be the best property design that will be safe and save time and energy? Planning needs to include where vehicles will be parked and where the house, other buildings and horse facilities will be situated in relation to each other for ease of use. You need to think about how you will divide up the land into pastures, and how the horses will move around the property. Which areas will be used most frequently by horses and humans? Consider separating horse and non-horse areas as this makes things much easier, particularly in terms of biosecurity. The relevant sections and the examples of property design later in this book will help you to make these important decisions.

Plan to be able to move horses quickly and safely around the property.

When planning a horse property, it is also essential to take horse behaviour into consideration; this helps enormously, both in terms of the wellbeing of your horses, as well as when moving the horses around the property. For example, any laneways that a herd of horses will be moving themselves along should be wide enough for them to traverse safely, with no 'dead ends' or sharp corners.

You may also find it helpful to refer to books on Permaculture design when planning your property. Even though horses do not fit neatly into a Permaculture system because they are not farm animals in the sense of being kept for meat, Permaculture design can still be applied to a horse property with many benefits. Basically, Permaculture is about creating harmony between landscapes, animals and people, increasing sustainability and reducing waste (including wasted effort). This fascinating subject cannot be covered in detail here, but there are many books on the subject and you can either buy them or borrow them from a library.

The safety and security factors

A property should be safe for all that live there including, of course, people (residents and visitors), horses, livestock, pets and wildlife. The design of the property should also take into account that inexperienced people (friends/family or house sitters) may need to take care of your animals when you are away. In ddition, the property should be laid out so that any visitors are kept away from the orses, both for their sake and the sake of your horses, at least until you are able ɔ supervise them. Horse facilities should be designed so that people can feed and necessary move horses around the property easily and safely. Inexperienced eople should never have to walk into an area such as a surfaced holding yard or table to feed a loose horse/s for example; even experienced people should avoid his situation.

• Facilities that are used to keep horses in a confined space, particularly if they are being 'group housed, need to be strong and there should be no protruding objects on any horse facilities that horses can bang into, get hung upon, trap any part of their body (hoof, head etc.) in or knock a human handler into.

Horse behaviour should be included in the design of facilities - for example designing tie up areas where horses can see other horses while being groomed and tacked up reduces potential injuries to people and horses, because the horses will be more relaxed.

By rounding off corners in areas that horses are kept together (such as paddocks and surfaced holding yards) and avoiding having gateways in the corners of paddocks, the risk of horses being cornered and kicked by other horses is much lower.

imple, well thought out designs such as these are no more difficult to achieve han unplanned tying areas and gateways, but can make life much simpler and afer for all concerned.

Typical behaviours that horses carry out that can lead to injuries include rubbing heir back, tail or belly, 'pawing' with a front leg, rubbing their neck/head and ushing/shoving each other. If horses are group housed and are having to ompete for feed, then aggression can occur (see Appendix: *Feeding confined orses*.

'ommon injuries that are caused by facilities include:

A trapped leg or hoof - this can result in the horse panicking and 'pulling back' which in turn can cause a serious injury. This injury usually occurs when a horse 'paws' at a gate, because the gate is preventing the horse from getting to where they want to go.

A trapped tail – again a horse will tend to panic and will usually leap forward when this happens, usually causing serious damage to their tail and even their spine. This injury can occur when a horse rubs their backside on something and their tail becomes trapped behind it. A further complication is that when a horse panics, they 'clamp' their tail down which can lead to it becoming more trapped.

161

- A trapped head – a horse can actually break their neck if they panic as a result of their head becoming trapped. This injury can occur when a horse rubs their head on a post that is too close to another upright, or not close enough to prevent a horse from putting their head between, or between any two uprights. It has been known for horses to trap their head between two trees growing close together or even in a hollow of a tree. They may also trap their head when they try to reach food by putting their head through a too small opening.

Almost anything is a potential hazard to a horse, all you can do is try to protect them with safe facilities.

- Deep cuts to the sides (flanks) are common when a horse brushes up against a protrusion. This injury can occur when there is a projection, such as a gate fastening that sticks out into the gateway, and the horse moves past it. People can be badly injured when leading a horse through a gateway that has a protrusion as they can be inadvertently knocked onto it by the horse.

Surfaces for horses (and humans) should be non-slip. Therefore:

- Concrete surfaces should be 'rough' not smooth.
- All-weather riding/training surfaces should have good traction; slipping can injure a rider/trainer if the horse falls.

If an all-weather riding/training surface is fenced, then the fence should be of a type that is not likely to cause injury to people or horses.

Reduce the opportunity for theft by careful planning. Aim to have the tack room near the house if possible. Keep in mind that a second driveway can decrease the security of the property; as can gateways that lead from paddocks directly out on to public roads. At the same time, the property must be easily accessible to the emergency services in cases of storm, fire and flood. All of the above points and more are covered in more detail in the relevant sections in this book.

The natural elements

Fire, flood and storms may be regular occurrences in the area that you live; therefore a property must be planned with these in mind. A property that is at risk of any of these natural disasters must have the relevant plans in place (e.g. a fire plan or a flood plan).

Fires

Fire can either come to the property in the form of a grass or bush fire, or it may originate on the property in the case of a stable, house or barn fire. Frequently, fire is something to which not enough thought is given until it is too late.

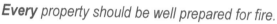

Every property should be well prepared for fire.

Horse properties are often located in the countryside where vegetation fires are common and are therefore at risk from fire on a regular basis. Plan to make the property as fire resistant as possible. This involves the correct placement of buildings and shelter belts, as well as keeping up with regular routine mainte-nance. It is imperative to develop a fire response plan in case of emergencies. Local fire services should be happy to give advice. There is also a link at the end of the section that will point you in the right direction for further help.

Find answers to the following questions from neighbours, the local fire service or the Bureau of Meteorology:

- Is a bush fire likely to occur in your area and, if so, when is the most dangerous period?

- How often does it tend to occur and is there usually a warning period?

- Does it follow a predictable pattern? Keep in mind though that fire is often very unpredictable.
- Which direction do the hot dry winds tend to come from? Fire usually comes from a specific direction; however you must be prepared for the chance that it may come from a different direction.
- Is the property/area part of a community fire reporting network?

If you are planning and developing a property from the beginning, you will be able to implement features that will make it safer if a fire occurs. An established property, on the other hand, may require some alterations to make it safer in the event of a fire.

When planning your property, think of it in terms of zones. The inner zone includes any dwellings, stables, surfaced holding yards and barns; the most valuable parts of the property and where people and animals are housed. This zone should be surrounded by a buffer zone of about 20m-30m (66ft-100ft). The outer zone includes the pastures/paddocks and treed/wooded areas. Draw up a plan from your property plan (see the section *Making a plan*) and identify the three different zones. This plan should include neighbouring properties also, as in order to reach your property, the fire may be travelling from them.

Any possible problems on the property should be identified and counteracted. This may mean moving something or creating something new. Fire risk problems include tall/dense vegetation, long grass, stacks of firewood, loose hay, certain types of trees and chemicals. Fire travels faster uphill, so beware of this. Pluses on a property include large bodies of water such as a farm dam/lake, a swimming pool, driveways and laneways (which can be used for fire breaks, evacuation routes and for fire service vehicles), gravelled areas, short grass (perennial pasture is both good summer feed and is fire resistant), a deciduous orchard, a vegetable garden, an arena/training yard (sand surfaced) and windbreaks of fire retardant trees. Plan to place fire breaks in the areas that fire is most likely to come from and reduce any fire risk problems in these areas.

One possible approach in an area that has a high grass/bushfire risk is to put a ring laneway/exercise track around the inner and buffer zone. This ring can serve various purposes in addition to being a firebreak, including use as an exercise/riding ring *and* being used to take horses out to the paddocks. Both sides or just one side of the laneway could be planted with fire resistant trees and could either be surfaced (an expensive option), or sown with hard wearing grasses that are able to cope with being kept short (lawn type grasses). It could then be strip/block grazed from time to time.

If fire or other natural disasters are a risk for your property you need to incorporate management strategies into your plans.

Drawbacks to this design are that the driveway to the house will need to cross this laneway, which will mean more gates to open and close when it is in use.

Alternatively, if the local authority requires that you have a permanent firebreak around the outside edge of the property a ring could be constructed there instead. Some local authorities in high grass/bushfire risk areas insist that a ploughed area is created around the edge of the property, but if you already have a well-managed firebreak in place you may be able to avoid ploughing. Ploughing will lead to weeds and possibly soil erosion. Both of these designs may have higher environmental maintenance issues because a track is more difficult to maintain than a paddock.

The Equicentral System design that we developed many years ago and now advocate (see Appendix: *The Equicentral System*) can work well in terms of fire safety. This total management system is a good way of keeping horses inside the inner zone in a *fire* situation.

Floods

Floods, like fires, tend to catch people unawares due to their rapid nature. Dry 'summer' creeks (e.g. in hot countries) can become a raging torrent in a matter of hours during a heavy rain event. People, horses and other animals can literally be swept away by floodwater, along with anything else in the water's path. Good property design should ensure that animals can get themselves to higher ground in a flood emergency, because you may not be able to get to them quickly enough and/or it may be too dangerous for you to do so. **The Equicentral System** (see Appendix: *The Equicentral System*) may work well for your property in this case. A gateway on the outer edge of the property, leading to the higher ground of a neighbour, may also be an option so that animals can easily be moved to higher ground in an emergency.

Storms

Storms bring high winds, lightning strikes, heavy rain and hail. Plan to have solid structures so that no there are no loose materials on the property that could blow around and injure people and horses in a storm. Loose metal roofing sheets are particularly hazardous in bad weather. Horses are often OK outside during a storm if they are used to adverse weather, but they should be given the choice of where to stand – outdoors or indoors. Horses are often unwilling to stand under trees during a storm; possibly because they innately know that branches may drop on them. Hail can be another factor during a storm and some form of shelter will protect them from that.

Floods, like fires, tend to catch people unawares due to their rapid nature. Horses must be able to get themselves to higher ground in a flood - you may not be able to get to them quickly enough.

—

See the **Equiculture website page www.equiculture.com.au/horses-and disasters.html** for more information about preparing for and protecting horses from disasters etc.

—

The aesthetic factors

What a property looks like is very important; it is a much nicer experience to live and work in a beautiful environment than an ugly one. Also, if you run a business from the property, it is very important that you present a professional image to your clients and potential clients. A poorly managed horse property also gives the horse industry in general a bad name. You do not need to have expensive, fancy facilities, just functional facilities that are kept neat and tidy.

Badly planned and managed horse properties are an eyesore and give other horse owners a bad name.

Also keep in mind that a property can either stand out and make a statement – white fences and ornate entrances do this, or it can blend in with the natural environment by having a low key approach. Fences and buildings in neutral colours are a good option in this case. The low key approach tends to be more sustainable and easier to maintain, no expensive and time consuming white

167

painted fences for example. Fashions come and go and certain facilities can end up looking dated over time.

Thankfully, good aesthetics and good environmental planning and management go hand in hand. A property with no trees, polluted water, weeds and bare overgrazed land is an eyesore, whereas clean water, trees and bushes and green pastures are very beautiful indeed. By taking care of the environmental planning and management of the property, the aesthetics will tend to take care of themselves. You will end up with a productive and efficient property which will by default look great and stand out from its neighbours even (or especially) those with fancy expensive facilities.

Thankfully, good aesthetics and good environmental planning and management go hand in hand.

Planning horse property infrastructure

Horse facilities can range from very basic to very elaborate. There are certain facilities that are essential on a horse property and others that are less so, depending on your needs, wants and budget. Even though many different facilities are described here, they are not all necessary on all horse properties. The most sensible thing to do is to start with the essential facilities and then add the others as and when money becomes available if you need them. The important point at the planning stage is to include in the plan any facilities that you would like to end up with so that, if and when funds become available, you have allocated a space for them.

A typical horse property will have some or all of the following features (these are not necessarily in order of importance):

● Areas for people and pets e.g. a house and garden.
● Access in, out and around the property e.g. the entrance, driveway and laneways.
● Areas for holding horses e.g. surfaced holding yards, shelters and stables.
● Areas for tack, feed and equipment storage, also manure storage.
● Areas for exercising and training horses e.g. an arena, training yard etc.
● Areas to graze horses - paddocks with pasture.
● A source and storage of water e.g. farm dams/lake, stream, river, well and tanks etc.
● Areas set aside for wildlife, shelterbelts, windbreaks and firebreaks.

On a small horse property space is at a premium, so in this case facilities should multipurpose if possible. An example of this would be a large surfaced holding yard that can be used for confining horses and for riding/training. On a larger property, space may not be as much of an issue, but it still makes good economic sense to design facilities that are multipurpose whenever possible.

The house and garden

If you have bought a 'bare block' you have the important decision to make of where to put the house. Sometimes this has already been decided for you to some extent as some sub-divisions already have an allocated 'building envelope' which you will have been told about when purchasing the block. There may also be covenants that restrict what style or size of house you can build. For example in some areas you may only get approval for a large house so that it will be in keeping with the surrounding houses. This is more common with properties that are in the 'urban fringe'. Alternatively, the covenant might stipulate that the house has to blend with the surroundings, for example in an environmentally sensitive area; therefore the colours of any buildings and fences may be restricted to neutral colours.

If you have no restrictions when it comes to deciding where to put the house on the block, there are several things to consider. A house at the back of or in the middle of the block generally looks nicer and is more private due to being further from the road. Driving up a long driveway, flanked by paddocks with green grass and horses gives a palatial feel to a property. A house in the middle of the block may also allow a panoramic view of the paddocks. There are however extra expenses involved when the house is situated further back on a block; a longer driveway and power/phone connection costs more (sometimes a lot more), so do your sums before deciding where to position the house. There are also issues to consider involving emergency evacuation in the event of a natural disaster – the nearer the house is to the public road, the easier it is to get off the property, and the easier it is for emergency services to get to the house if needs be.

It will be safer in terms of security of horses and equipment if the house is positioned at the front of the block (near the road), with the horses and facilities behind it. In this case, the sheds and equipment can be situated behind the house away from general view. With this option you do not need a gate that has to be opened and closed every time anyone enters or leaves the property. Instead, a gate that is always kept closed and leads to the horses and facilities and other equipment can be situated at the side of or behind the house. People (visitors etc. can arrive and leave the property without any danger of them leaving open a gate that might mean animals getting out on to the road. Another plus for this option is that it will be cheaper in terms of driveway construction and power/phone connection

Aim for low maintenance so that you can maximise your time spent with the horses rather than carrying out maintenance work - white painted wooden fences are high maintenance.

The house and other buildings need to be angled the best way for maximum sun in winter and maximum shade and breeze in summer, protection from the wind and to take advantage of the views etc. Before deciding, you also need to see where the water flows in wet periods. To find out what the weather patterns are on a property, it is helpful if you can live on the property for a complete year before building. This may or may not be possible, for example in some areas you may be allowed to live on the land in a caravan while developing the property. If you cannot, you need to speak with locals to find out if there is anything you need to know about before deciding, such as which direction the bad weather comes from at certain times of the year. Your builder should also be a good source of information about how to position the house.

Buildings should not be positioned too close to a boundary, as this can create problems if you need to improve drainage or for maintenance at a later date. Another factor is that if your house is close to a neighbour's house and they are not diligent about fire preparedness, you will not be able to control this aspect and what they do or don't do may affect you and your family. Building regulations vary from area to area, but a gap of several metres between the boundary and any buildings is usually required in rural areas. It is a good idea to aim for much more than this. Keep in mind that previously (while in suburbia) being too close to neighbours is one of the reasons that people want to move out to a more rural area, so don't create future problems by building too close to neighbours if you can avoid it.

The house can either be located in the same area as the horse facilities or separate to them. There are many advantages to having them close, including the reduction in costs for supplying power to both, added security for the horse facilities due to proximity to the house and ease of use travelling between the two when caring for horses. There are also advantages in terms of horse safety in the case of natural disasters such as fire and flood (see the section *Making a plan*). The disadvantages are possible smells and flies, however if the property is managed well, these can be minimised. The stables (if building them) or covered surfaced holding yards can be situated downwind of the house to reduce these issues, but good stable or surfaced holding yard management should make this unnecessary. If the property is to be a commercial horse property though, there may be other reasons for wanting the house away from the horse facilities, so that home life and business life are separated.

When deciding on how large the garden is going to be, think about the following points. Large, ornate gardens take time – and lots of it – to manage. Avoid having large 'lawns' that require frequent mowing and lots of water to keep them looking good. Instead, plan to have more space for paddocks. With correct pasture management, your paddocks will look as good as any garden and, instead of having to spend hours cutting grass with your ride on mower, your equine 'ride on mowers' can do the work for you in the paddocks! Your powered ride on mower will then only be required for tidying paddocks as and when horses are rotated on to the next paddock. This is a lot less work than maintaining large 'lawns', more sustainable and far more satisfying.

The house and other buildings need to be angled the best way for maximum sun in winter and maximum shade and breeze in summer, protection from the wind and to take advantage of the views etc.

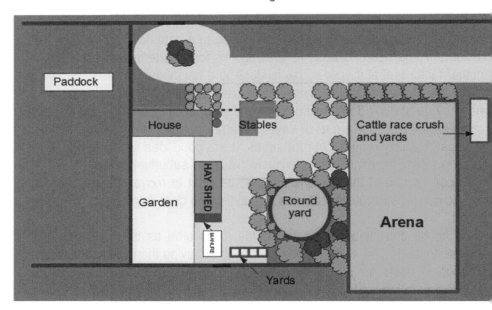

Make sure you build good enclosures for any pets such as dogs and cats. On a commercial property, dogs should not be allowed to be loose while clients are on the property. Even on a private use property, you will need a safe secure area for your dog/s. Dog/mesh fencing around the garden works well. Offset electric fencing can be used to keep horses away from it if this fence is also part of a paddock fence. Never feed horses or allow visitors to feed horses over the garden fence, or they will hang in this area for large periods of time and put pressure on the fence, which may also result in fence injuries. Cats should be enclosed at night when native nocturnal animals need to move around the property to find food. Cats do best controlling rats and mice on a horse property and therefore reducing food for snakes if they are an issue, but you must make sure that they are not also annihilating the local wildlife.

With correct pasture management, your paddocks will look as good as any garden.

Property access

The property must be easy and safe to enter and exit. Plan to put the entrance gateway at least the length of a vehicle and horse trailer in from the fence line at the front of the property so that a vehicle will not have to be stationary in the road when the driver or passenger are opening and closing the gate. This will also make riding a horse in and out of the property much safer if that is what you plan to do.

Make sure that the exit from the property is in a position where you can see in both directions of the road. Remember that if you are towing a trailer, you will need more time to exit than you would without one. Likewise, turning into the property when towing will also require slower manoeuvring

A horse property should always have a closed gate preventing any escaped/loose horses from getting out on to the road. Depending on the layout of the property, it may be possible to place this gateway beyond the house and before the horse facilities/paddocks, eliminating the need to open and close a gate when entering and leaving the property.

You may have considered installing a **cattle grid** at the entrance of a property to reduce having to open and close a gate when entering or leaving the property. There are some things to keep in mind with cattle grids. Horses are *much* more flighty than cattle, and a panicking horse will *not* stop at a cattle grid. A panicked horse will either jump cleanly over the grid, in which case the horse is then out on a public road, or aims to jump and gets a leg caught in the bars, often wit disastrous consequences. During a flood, horses will not be able to see a cattl grid and can get trapped and/or injured. Some horses have also been known t learn to 'tip toe' over a cattle grid, again resulting in them getting out on to th public roads. Cattle grids are not cheap to buy or install, so if you want to avoi having to physically open and close a gate, consider installing a solar electric gat instead.

If the property has a longer driveway because the house/horse facilities are s€ in the middle or at the back of the property, the front gate can be positioned at th house/facilities end of the driveway. This means that vehicles can be well awa from the road when the driver/passenger is opening or closing the gate and als means that it is easier for people living in the house to see and check that the fror gate is always kept closed.

Wherever the gate is to be positioned, try to make sure that it is not positione on a slope; towing and large vehicles such as trucks carrying horses/hay etc. ar in more danger of rolling forwards or backwards than ordinary cars due to the weight, and some visitors may have not have a good enough handbrake to hol their vehicle if the driver has to get out of the vehicle to open/close the gat€ Setting off from a standing start on a sloping gravelled driveway can also b difficult or impossible with a heavy vehicle.

This driveway has the gate positioned well up the driveway which makes it easier to enter and exit with a large vehicle.

The driveway should not be rutted or unduly steep. Steep driveways cost more to maintain due to the damage that water runoff and gravity creates and they are more difficult for vehicles such as cars towing trailers and trucks to traverse. So if possible, plan to avoid a steep driveway, or plan to have a separate area that larger vehicles can use for access and have the steeper driveway leading to the house only.

If the driveway is to be long, it can be positioned to go up one side or through the middle of the property. The latter may involve more fencing, but usually looks nicer. You may even plan to have separate driveways for the horse facilities and the house, although on a small horse property, this will use a lot of space that could be used for paddocks etc. This arrangement is more important if the property is to be commercial e.g. a riding school or livery yard. Make sure you plan to have lots of room to turn vehicles (cars towing trailers and trucks) around at the facilities end of the driveway. Be generous in your allowance for the width of the driveway. Either side of the driveway can be planted with bushes or trees to act as a windbreak. Make sure that you leave enough space so that when these plants mature, there will still be room for large vehicles to traverse the driveway. Trucks need at least three metres width, more if there are curves or turns to negotiate.

If the property is in a flood risk area, make sure the driveway is constructed to protect you, your family and your animals from being unduly stranded on the property during flooding. This will mean avoiding any low laying areas on the property.

Laneways

Laneways mean that horses (and vehicles) can be moved around the property quickly and safely. For example it is not safe to lead a horse through a paddock that contains loose horses. Laneways should link paddocks to the main facilities area for convenience. If the horses live in herds, they can be moved around 'en masse', when necessary via the laneways or they can even move themselves if you incorporate **The Equicentral System** that we advocate (see Appendix: *The Equicentral System*).

These laneways can serve several purposes; laneways can double up as exercise tracks in some instances and can be planted with fodder trees on one or both sides. As well as providing fodder, these plantations will be a windbreak and will give shade and habitat for wildlife. You need to get local advice about what would be the best trees to plant in your area.

The best location for laneways is on ridge lines where the ground will be dryer; if this is not possible, avoid low lying wet areas. If a laneway is created along a contour, it can help to divert water along the slope so that it slows the speed of water running down the hillside. Laneways can also be grazed from time to time by quiet horses.

If laneways traverse a hillside, they should not have too much sideways slope on them; a horse can slip and fall easily when walking across a slope. Keep in mind that if the laneway fences were not there, the horses would be able to pick their own way down the hill. The fences will direct them *along* the side of the slope more than they would choose to do voluntarily.

Laneways mean that horses (and vehicles) can be moved around the property quickly and safely.

On smaller properties, it is not usually necessary to be able to drive a vehicle on the laneways except in an emergency and for property maintenance. If laneways receive lots of use and/or you live in a wetter climate, they may need to be surfaced with some form of crushed rock etc. Look out for local roadworks; the old road surface can make an excellent (and often free) base material for a laneway.

Bearing all of this in mind, plan to *minimise* laneways. If the paddocks are able to connect to the horse facilities without needing laneways, this will save fencing, surfacing and maintenance; laneways are harder to maintain than open paddocks. So if a property is flatter, the paddocks may be able to 'fan' out from the horse facilities for example (see Appendix: **Minimising laneways**).

Horse facility positioning

When deciding where to put the horse holding facilities on your property, consider such factors as distance from the house and other buildings, your neighbours, climatic factors, distance from any waterways, drainage and the location of power.

Generally, it is preferable for any horse holding facilities to be reasonably near the house. This way it is possible to check easily on horses late at night when necessary (see the section *The house and garden* for more pros and cons).

Tack rooms and feed rooms are best if separate from each other, as the latter create a lot of dust. Large amounts of hay should not be stored near a house, stables or surfaced holding yards due to fire risks.

Remember to take your neighbours into consideration when planning where to put horse facilities; neighbours will understandably not usually appreciate horse facilities being built near their house.

Stables and surfaced holding yards should be downwind of any residences due to flies and smell, although these should not be a problem on a well-managed property.

Depending on where you live and whether sunlight or shade or protection from strong winds is the priority will determine how buildings such as stables and shelters should be positioned. In tropical climates, it is more important that breezes pass by or through any buildings during summer. During winter in tropical climates, temperature is not really an issue for horses, horses thrive in cold weather and even though owners wrap horses up thinking their horses need the protection of blankets and rugs etc. they generally don't. However, in such a climate, shelter from heavy seasonal rains will be important. In temperate (colder) climates, protection from the winter elements such as wind and rain is more important, along with some shade in summer. Even though horses can usually cope with colder temperature extremes than most owners give them credit for, domestic horses, unlike their wild relatives, are unable to seek a more sheltered environment other than what is provided by an owner. Therefore, it is important that they can get themselves to a shelter at all times.

Holding areas should be located in elevated areas of the property, high enough to safely hold horses in the event of a flood. Buildings should be 20cm–30cm (about 9ins-1ft) above the outside ground level and *at least* 2m (6ft 6ins) above the wettest water level for your area. Ideally there should be a 2-6% slope away from buildings for surface drainage. A water diversion ditch should be dug around the back of any buildings if they are on a hill.

Holding areas should be positioned *at least* 50m-100m (160ft–320ft) away from water courses such as streams and rivers, but may need to be much further if in a flood prone area. Horses can be trapped in facilities if there is a flood and contaminants such as raw manure and urine may regularly get into the waterway if they are positioned too near. If there is a chance of contaminants from manure and urine seeping through the ground surface down to the water table, then stables and surfaced holding yards should be sited on an impervious layer such as concrete or compacted limestone. Speak to your local authority or agricultural department for more advice.

Depending on where you live and whether sunlight or shade or protection from strong winds is the priority will determine how buildings such as stables and shelters should be positioned.

Have the subsoil evaluated before building; sand and gravel are better than clay for buildings because clay tends to move when wet and shrink when dry. If the subsoil is mainly clay, it may be necessary to have the area excavated and filled with rocks and road base or limestone. This will need to be rammed (compacted) by heavy machinery, or be left to settle for several months or longer before building

Shade and shelter positioning

A shelter should be built in a high and dry area whenever possible and placed to take account of hot summer winds, cold wet winter winds etc. It should be placed either right up to a fence, or well away from a fence to avoid a horse becoming

apped between the fence and the shelter; placing it away from a fence means
‫at the back of the building can also provide shelter at different times of the
ay/year. There are also a few things to consider, depending on whether the
‫helter is for an individual horse or for a group of horses.

‫dividual shelters

horses *are* to be kept separately, then they each need a shelter. Aim to position
‫em where the horses will use them rather than ignore them. **Some other things**
‫ think about are:

In this kind of set up horses spend a lot of their time standing at the gate -
waiting for supplementary feed and/or standing on either side of the fence from
each other.

Companionship is so important to horses that they will forgo comfort and even
food to some extent in order to be near each other.

If you ignore this fact about horse behaviour and put the shelters where *you*
think they need to go, the horses will mainly ignore them. Therefore, shelters
should be where the horses tend to stand, not where you would *like* the horses
to stand (see the section **Domestic pastured horse behaviour** in **The
Equicentral System Series Book 1 – Horse Ownership Responsible
Sustainable Ethical** about why domestic horses behave the way they do when
in a paddock).

*If you ignore this fact about horse behaviour and put the shelters where you
think they need to go, the horses will mainly ignore them.*

Therefore, the entrances to individual holding yards or paddocks is usually th best spot for shelters so that the horses can see and preferably touch each othe while using them. One roof can then service two areas.

A fence can be put around the individual shelters to create smaller holdin yards within the larger surfaced holding yards/paddocks; if the area is grassed will need a surface. These areas will then become useful areas for handlin feeding and confining horses when necessary (if the land is too wet, too dr overnight to eat hay etc.).

Alternatively, depending on the layout of the surfaced holding yards/paddock: one large roof can be built over the corners of four surfaced holding yards/sm paddocks so that four horses can stand together while in the shelter. This kind shelter can have either open or filled-in sides depending on the win direction/climate. The downside of this arrangement is that the holding yards a then not near the entrance, making supplementary feeding etc. more difficult. solution is to create a laneway which leads into this covered area.

A fence can be put around the individual shelters to create smaller holding yards within the larger surfaced holding yards/paddocks.

Shelters for groups of horses

If you already have paddock shelters in individual holding yards/paddocks an they are not positioned correctly, you have a couple of options. You coul dismantle them and rebuild them in a more suitable spot or, alternatively, you ma be able to improve the shelters so that horses find them more attractive k removing or rearranging certain walls so that the horses can at least see eac other while using them. Depending on the distance between them, you may b able to erect a shade sail between the two solid structures in order to make them more 'user friendly'. Plan to also enclose them in a small surfaced holding yard for the reasons outlined above.

Alternatively, depending on the layout of the surfaced holding yards/paddocks, one large roof can be built over the corners of four surfaced holding yards/small paddocks.

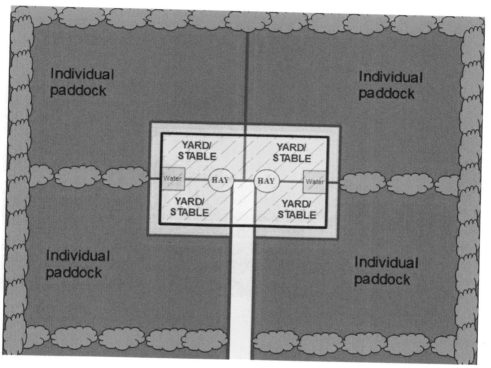

If you keep your horses in a herd, the shelter needs to be large enough for all of them to access it at the same time so that no horse is excluded from the benefits of it. This can be difficult as you may need it to be larger at some times than others – so you need to plan for all eventualities and make it larger rather than smaller.

Some other things to think about are:

- If you are rotating the herd around the property in accordance with a rational grazing system for good pasture management, you will need a shelter in each paddock.

- Building a shelter of that size in every paddock makes little economic sense as each large and expensive shelter will only be used for part of each year due to the amount of time that each paddock will spend in the rest and recuperation stage – as part of your rotational grazing system for good pasture management.

- Aim to minimise shelters by having paddocks leading to **one** large shelter. This should be in a surfaced holding yard (and should have a water trough). The whole herd can then access this area from whichever paddock they are currently grazing; the paddock should always be open when the horses are grazing so that they can always access this area. See Appendix: *The Equicentral System*.

*Aim to minimise shelters by having paddocks leading to **one** large shelter that is situated in a large surfaced holding yard.*

Fence and gateway positioning

When re-fencing a property it is usually a good idea to construct the fences in two stages. The boundary fences are a priority and should be improved ASAP if they are not already safe. The internal fences are not as important and can come later; even a year or two later, meanwhile temporary electric fencing can be used.

Boundary (perimeter) and internal fences serve different purposes on a horse property to a large extent. The boundary fence is in many ways the most important fence on a property as it keeps any horses (and any other animals) on the land and can also help to keep intruders out (depending on the construction). For this reason, it should be the strongest fence on the property.

On the other hand, the internal fences are there to keep horses in the correct areas within a property, but do not necessarily need to be as physically strong as a perimeter fence.

Boundary fences

Before erecting any new boundary fencing, check where the boundary actually is (find the survey posts) so that you do not end up in a dispute with your neighbours – which may result in having to redo fences later on. There is usually no flexibility concerning the placement of the boundary fence. Sometimes if there is a problematic section of land on your boundary (e.g. a very steep and rocky area) you may decide to fence this area out of your property if it is easier to do so. This area will then become land for wildlife. Keep in mind that you will still be legally responsible for this land and anything growing on it.

Another area of potential dispute with your neighbours is the subject of who pays for what with a shared boundary fence. The cost of simple types of fencing is traditionally shared equally between neighbours in rural areas however, if you want something more elaborate and expensive, you will be expected to pay. Your neighbours may have no interest in upgrading the fences if they have no livestock or a different type of livestock to you, so in this case you will need to stand the costs yourself.

Before erecting any new boundary fencing, check where the boundary actually is.

If you have a river on your boundary, the perimeter fence may fence this area out of your paddocks so that large grazing animals do not disturb the ecology of the river. There is often funding available for a fence that runs along a waterway so check before construction.

It is not recommended to only use temporary electric tape or braid or even fine electrified wire for a boundary fence. However these materials, combined with more permanent materials such as galvanised plain wire, 'horse mesh', timber etc. make a highly suitable boundary fence. Keep in mind that if electric is incorporated into a boundary fence, you may be required to display warning signs for the public (check local regulations).

Internal fences

The placement of internal fences requires good planning if the land is to be managed well. If there are not enough paddocks, horses will spend too long in them and over graze certain areas while ignoring others. If there are lots of small paddocks, you will be able to keep horses moving around the land (rotating your grazing), but small paddocks cost more to set up as more fencing is required and smaller paddocks are also harder to maintain due to the difficulty of getting machinery into and around them.

*Lots of small permanently fenced areas **do not** allow for versatility.*

If you have just moved on to a new property, it is usually best to put off building permanent internal fences until you have had plenty of time to plan where you want them. Leaving the permanent internal fences until later has many benefits. It enables you to carry out pasture improvement without permanent fences getting in the way. It allows you more time to plan where the internal fences need to go; there are many factors to take into consideration - you can try out different configurations before committing to permanent fence positions. It also saves money that can be spent on pasture improvements which will in turn save you money on feed.

In fact, there is no reason why the internal fences cannot always be portable electric fences – this works well on a small horse property in particular. Portable electric fences allow the layout to be changed easily. This then negates some of the problems described above of having too many small paddocks, because they can be easily removed if necessary etc. An electric fence works well as an internal fence as long as it is properly constructed and maintained.

The shape of the paddocks will be dictated by the available space (after space for any buildings has been allocated) and the topography (contours, features and shape of the landscape). There are many things to take into consideration such as changes in land type, hills (undulations), waterways, vegetation such as any existing trees/bushes and any planned plantations, and how you plan to manage your land and horses; your management system.

Wet areas and dry areas must be separated into separate paddocks so that they can be grazed at the appropriate time of year; wet areas should only be grazed during the dry times of the year and vice versa. Square paddocks are the most economical to fence, but they are unlikely to fit in with designing your property to minimise land degradation if the land is undulating.

Poor subdivision of your land can be the start of problems such as soil erosion. Whenever possible, aim to fence along the contour lines in order to reduce erosion. A contour is a line that connects points of equal height on the land. Ridgelines and valleys can be used as a starting point.

The fences themselves do not stop soil and water from travelling down a hillside, but incorrectly placed fences force animals to track up and down a slope. This creates channels which water will use to travel down the hillside, quickly speeding up the erosion process. By fencing along contours, any tracking lines created will be across the contour – which will slow the flow of water. Placing the fence along the contour lines also means that trees and bushes can be planted on this line, for example on the other side of the fence in a laneway that separates two paddocks, further slowing the water.

Where there is a greater slope, it may be necessary to create a bank of earth on the top side of the laneway to help divert and slow water. Any cultivation to improve drainage and water uptake such as subsoil ripping can be done using the fence line as a guide. When fences have to go up or down a steeper slope, aim to make these sections as short as possible.

Never have acute angles and even avoid right angle corners in paddocks if horses are kept in groups. Horses can be 'cornered' in them while playing. Horses also like to stand in corners staring off into the distance, which results in these areas becoming bare, compacted and degraded. Paddocks with rounded corners are safer for horses and easier to maintain. Fast moving horses are guided around

185

the corner rather than into it and a tractor pulling a pasture harrow or mower can get up to the edge of a rounded corner. Acute and right angle corners in paddock that are already fenced can be eliminated easily by fencing across the corner with electric tape, braid or wire. This fenced off area can then be planted with bushes or trees, thus creating habitat for wildlife whilst making the corner safer and easier to maintain.

Incorrectly placed fences force animals to track up and down a slope.

It may be helpful to think about organic shapes for paddocks on your property, remembering that nature rarely has straight lines.

The paddocks can have fences that follow the natural lines of the landscape avoiding tight corners. These types of fences are more in harmony with their surroundings, but will be harder to tension if using plain steel wire. This is another reason for using electric fencing for internal fences, you can create as many turns and curves as you like.

Plan paddocks so that there are no power line/telephone line guy wires or other potentially dangerous permanent fixtures inside them. If this is not possible, fence around the unsafe object within the paddock. Remember, horses can and will trap parts of themselves or injure themselves on any dangerous obstacle.

It may be helpful to think about organic shapes for paddocks on your property, remembering that nature rarely has straight lines.

Gateway positioning

Gates and gateways are potentially the most dangerous section of a fence for various reasons. Gateways are areas of high activity on a horse property. Horses tend to congregate in gateways (waiting to be let in for feed) and can crowd each other in this tight spot; also horses come into close contact with the gate when passing through (which is a potential for injury). This is also an area that people and horses are in close contact with each other so many injuries to people occur in gateways as well as horses, see the section **Gates and gateway safety in particular**.

There are several things to keep in mind when it comes to gateway positioning:

- Ideally, a gateway should be positioned so that if horses accidentally get out through a gateway, they are still contained on the property.

- Avoid placing gates in a position that make it possible for someone to let your horses out easily e.g. on to a road; any gates in this position should be locked so that horses cannot be let out either inadvertently or deliberately. In a rural area, this is not usually as much of an issue as it is in a more built up area, but it is still better to be safe than sorry.

- Avoid positioning a gate in the corner of a paddock for the reasons described in the section **Gates and gateway safety in particular.** If a gateway has to be in a corner, consider building a surfaced holding yard around the gateway that keeps the horses back from the corner.

- Also, avoid positioning a gateway in a wet, low-lying area, as this will result in even muddier gateways and lead to erosion when the mud dries out and blows away as dust.
- Consider putting additional gates in boundary fences between your property and the neighbour's property in case you need to evacuate your horses in the event of flood or fire. Think of these as your 'emergency exits'. These will obviously work both ways, allowing your neighbours the same benefits. This is preferable to gates (emergency exit gates that is) that lead directly out on to public roads from paddocks. This may mean that in some cases a paddock has two gates; one for everyday use that is in the most logical position for this purpose, and one that allows the animals to get to higher ground in the event of a flood. These 'external' gates may also come in useful if you end up 'borrowing' a neighbours animals (e.g. cows/sheep) for cross-grazing purposes.

There are various solutions for gateway problems.

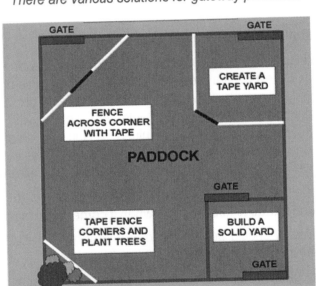

- If horses are going to be allowed to bring themselves back through gateways (such as when using **The Equicentral System** that we advocate - see Appendix: **The Equicentral System**), they need to be positioned so that horses do not have to walk too far *away* from their intended direction of travel. If you are utilising **The Equicentral System**, gates *can* be in the corner of a paddock because in this case, the horses are never fastened behind them. If the gates are positioned incorrectly, the horses will struggle with the concept of having to walk in the opposite direction to where they want/need to go and can ge 'trapped' in the corner of a paddock. If horses get separated because some horses get it right and some don't then this can lead to separation anxiety and its resultant chaos. A good place for a gate then, whether the horses are moving themselves or you are moving them, is just in (a few feet/metres) from the corner that is nearest to where the horses are going. The actual corner can be rounded off and vegetation planted in it, making the gateway safer and providing land for wildlife at the same time.

*The gate needs to be positioned so that the horses **do not** have to walk away from their intended direction of travel. So for paddock A there is a correct and an incorrect position for the gateway.*

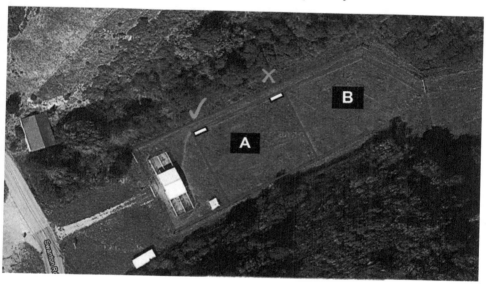

All-weather surface positioning

Whenever possible, areas that are used for working horses should be situated so that they can be seen from the house or a place where other people are working or both. This is because people who are handling and working horses should always be able to be seen by other people safety reasons. On a busy horse property, this may be near the other horse facilities where there are other people about (staff or other clients for example). On a private property, the area may need to be near the house so that other family members can periodically check on those working horses. This does not necessarily mean that they have to be in full view of the house, just somewhere that can be seen from certain parts of the house if necessary.

Aim to position an arena or training yard where it can be seen from the house or stables - somewhere that other people are able to keep an eye on people working horses.

An all-weather surface can be positioned on a site that would otherwise not be desirable for other uses, for example a sloping area that is unsuitable for use as a paddock. A sloping site will help water to drain away. Keep in mind though that earthworks for a sloping site will cost quite a bit more than for a flat site.

High ground is generally better than low-lying ground but if an all-weather surface is correctly constructed in terms of drainage, it can actually be positioned almost anywhere on a property except a flood zone. It can be a good idea to have a soil survey done before beginning work or even deciding on a site; some soil types are better than others, with some draining well naturally and others not so well. Generally speaking, sandy loam soils are better than clay which can be more problematic in all-weather surface construction. Excavation can help by removing more of a problem type of subsoil and then better base materials can be trucked in. Of course this will cost more. If you have a potentially problematic area to deal with then getting professional advice and help will be even more important.

If rain is a problem, position the all-weather surface so that the sun will help dry it out. If too much sun is a problem, think about whether you can use any natural shade.

A sand covered outdoor all-weather surface can also be used as a fire buffer zone in areas prone to bushfires, so you may want to position it between the house and the area that fire is likely to arrive from. Trees and other vegetation can be planted strategically to reduce dust and to create shade for the riding/training area (these need to be fire resistant) and act as a windbreak if necessary.

A sand covered arena can also be used as a fire buffer zone in areas prone to bushfires.

Preferably, riding/training areas should also be near the tack room so that if you need extra equipment while training or riding it is not too far away. Alternatively, a small shed can be positioned on the edge of the area for gear items (such as lunging equipment) that are frequently needed. Make sure that trucks will be able to access the area even after completion for topping up the surface in the future.

Training yard positioning

The considerations for where to put a training yard are the same as for a riding arena. In addition, because a training yard is generally smaller than a riding arena, it can be placed in an area that might be too small for a riding arena. However, it can be useful for a training yard to link to surfaced holding yards and/or to link to the riding arena if there is one. There are various benefits to having a training yard and arena near each other. You might, for example, want to start off working your horse in a smaller area when you first mount, but move into a larger area once warmed up. Remember that a training yard (or riding arena) can also double up as a surfaced holding yard if required.

191

This arena and round yard are cut into a slope and are positioned near other horse facilities and the house (off to the right bottom of photo). The large sand area also provides a fire buffer between the house and the surrounding bushland.

Manure management planning

The following information is an excerpt from the second book in this series *The Equicentral System Series Book 2 - Healthy Land, Healthy Pasture, Healthy Horses* where there is much more information about manure including - **the 'dunging behaviour' of horses, 'horse-sick' pasture, parasitic worms**. **Manure management strategies** including - picking up manure in paddocks (pros and cons), using dung beetles and other insects. **Improving manure strategies** including - composting manure, using composting worms (vermicomposting), using chickens. Lastly, **what to do with** improved manure and **managing** surplus manure.

Obviously manure must be collected from areas where horses are confined (such as stables and surfaced holding yards), and some horse owners also pick up manure from paddocks. This collected manure needs a place to be stored and you may decide that you would like to compost this collected manure.

A three bay composting system. Covering the pile will help it to stay moist without allowing rainwater to get in and make the pile too wet.

Composting manure

Composting is a process whereby bacteria and fungi consume oxygen while feeding on organic matter. This results in the release of heat, carbon dioxide and water vapor into the air.

Decomposition of manure starts as soon as it is passed and the rate of decomposition depends on handling and storage methods. Composting manure is a great way to turn a nuisance into an asset, it is a method of *speeding up* the process of decomposition that occurs naturally to organic materials; organic materials can be defined as something that was once alive. By providing the correct conditions, the micro-organisms that decompose organic matter, such as manure, are able to work to their full extent. The end result is a far superior product than fresh manure.

This is a blown air composting system. Air is blown through the manure pile helping it to decompose.

Composting bays

Bays should be built for efficient composting as, unless a pile is compact, it does not reach the required temperature to compost properly. A pile should be at least 1m (3ft) high to reach high enough temperatures and 1.5sqm to 2sqm (5sqft to 6.5sqft) in area. A compost heap can either be purpose built or made simply from materials such as old hay or straw bales. These bales can be periodically mixed in to the heap, and new ones used to start another compost heap. Other materials that can be used to build a temporary or permanent compost heap include wood pallets, bricks, old tyres, concrete blocks, railway sleepers, steel sheets, timber, concrete etc.

The amount of space required to compost your manure depends on how many horses you have and how they are managed. Other considerations include whether you have access to machinery, or you will you be moving the compost around by hand. A 10m by 10m (33ft by 33ft) pad will comfortably house three compost bays with room to move machinery. This area is large enough to cope with the manure from one to five horses that are kept confined on a regular basis (e.g. in overnight, out through the day). Three bays are ideal so that one can be filling, another composting and the other full of finished composted material that is available to use. A smaller area will be fine for a horse property where less manure is collected (e.g. the horses are at pasture for much of the time) and or machinery is not used in the process. It may be a good idea is to start with a smaller area and add to it if the need arises later on. Make sure you position the pad with this in mind.

This steel sided composting bay also has a perforated pipe that allows air into the manure pile.

A compost heap should be located well away from a water course. There should be a buffer zone of vegetation between it and any waterways. It should also be in the sun. Check with your local authority about regulations for your area. Try to locate storage sites so that filling and emptying the bays will be convenient. It may be possible, to locate a compost bay below the level of the stables so that manure can be tipped into the bay without having to fork it out of the barrow. Covering the heap (with a tarp or piece of old carpet) keeps it moist, reduces the breeding ground for flies and reduces loss of nitrogen to the atmosphere in the form of ammonia. It also prevents water from running through the heap which can contribute to unwanted nutrients entering the waterways.

The ground surface should be non-permeable to prevent leakage of manure into the soil which can then get into the waterways. If possible, a concrete pad should be laid. This is even more important if machinery will be used to shift the compost otherwise the area will quickly become waterlogged and smelly.

Grade the surrounding area to keep surface water from running over or through the manure and then into farm dams, streams or rivers (taking pollutants with it).

—

The subject of manure and manure management is a very important subject with regards a horse property so make sure you read the extensive information in *The Equicentral System Series Book 2 - Healthy Land, Healthy Pasture, Healthy Horses*. We also have some information on the **Equiculture website www.equiculture.com.au/horses-and-manure.html**

—

Water management planning

The following information is an excerpt from the second book in this series *The Equicentral System Series Book 2 - Healthy Land, Healthy Pasture, Healthy Horses* where there is more information about horses and water including - **creating and caring for a riparian zone**, **conserving water** and **water problems**.

Sources of water

Water primarily arrives on your land via rain that falls on it directly, and as rainwater running neighbouring land, which travels across your land towards a body of water such as a farm dam, lake, wetland, stream or river.

Other potential water sources on a property include:

Mains water, which is water that is piped to your property by the local authority in urban and some semi-urban areas – but not usually rural areas depending on which country you live in.

Well (bore) water, which is water that is pumped up from the underground water table (aquifer).

Natural spring water, which is water that naturally rises to the surface (from the underground water table (aquifer) in certain areas.

Streams or rivers that run through or alongside the property.

This photo may look idyllic at first glance, but look at the bare soil and therefore lack of riperian area around this waterway.

Mains water

Different countries have different arrangements for 'mains water'. In the UK for example, most properties are connected to the mains with only 1% not being connected. Therefore, even many rural properties have mains water. By contrast in Australia and New Zealand, rural properties rarely, if ever, have mains water. Only those that are near to suburbia tend to have it. In the USA and Canada, similar situation occurs with 15% and 25% of homes respectively not having mains water.

If a horse property *does* have mains water, then it may be metered and an added expense to use for watering stock. Consider installing rainwater storage tanks on any buildings that can then be used for watering animals. These are becoming more popular even in countries that do not traditionally use them (see the section **Water storage tanks**). Ideally, a property would border a running waterway and have permission to pump that water.

Well (bore) water

Well water is water is drawn up from a natural water aquifer underground, often called the 'water table'. In modern times, a 'bore' hole is drilled until underground water is found and this water is pumped up to the surface by an electric pump. Before electric pumps were used, water was drawn up to the surface physically hence the old fashioned image of a well being a covered circular wall around hole into the ground and a winding mechanism for the bucket.

In some areas, the water table is huge and contains abundant high-quality water. In other areas, it is either non-existent, limited or of poor quality. Worldwide demand for water is enormous and in many areas these underground stocks water are dwindling through overuse. Be aware that well water might contain minerals that can be harmful to stock and fauna. It can be very high in salt particular and can therefore cause a lot of damage to the land. If you have a well on your property always have its water analysed before use. Seek advice from the relevant local authorities before constructing a new one.

Aquifers vary in how close they are to the surface. Well water is exhaustible and in some areas it is being used faster than nature can replace it so causing new environmental issues. If your property does not have a well but you are thinking putting one in, you need to weigh up the costs with the benefits. In many areas there are now restrictions on putting down both domestic and irrigation wells.

Natural spring water

To have spring water is obviously highly desirable as it should mean that the property will always have water. Some spring water can be quite heavily mineral ised, which may make it unpalatable at least initially until the animals become accustomed to the taste. Like well water, it should be tested before using for stock. A farm dam or lake may be able to be created around a natural spring which results in a constantly full waterway.

Streams and rivers

Domestic animals should not have direct access to streams and rivers for use as drinking water due to the extensive damage that they can cause to the ecology of the waterway. Also, if a waterway has a sandy base, the horses will take up this sand each time they drink, potentially leading to sand colic over time due to an overload of sand in the gut. In addition, water that is not moving becomes stagnant and this water is then not clean enough to drink. If the property has a licence to harvest water from the waterway, it is better to pump the water to one or more holding tanks that are placed higher up on the property (called a header tank – see the section **Water storage tanks**). Gravity will then allow this water to be reticulated to water troughs below this level.

*Domestic animals **should not** have direct access to streams and rivers for use as drinking water due to the extensive damage that they can cause to the ecology of the waterway.*

Storage of water

All properties have the opportunity to collect rain water both directly and as run off. If you live in an area that tends to run out of water at certain times of each year, then capturing this water and storing it in tanks and farm dams for later usage is an efficient way of ensuring water self-sufficiency.

Water storage tanks

Rain water from the roofs of buildings can be collected and stored by fitting gutters and spouts to all buildings that run to water storage tanks. Water storage tanks can be concrete, corrugated steel or polyurethane. These tanks are very common on rural properties in Australia and New Zealand, but are only just becoming available in other countries. If you live outside Australia or New Zealand and you are interested in harvesting and storing water sustainably, do some research and see if they would prove to be cost effective in your locality. To get started, do an Internet search for **Water storage tanks**.

Rain water from the roofs of buildings can be collected and stored by fitting gutters and spouts to all buildings that run to water storage tanks.

Any overflow/excess water should be channelled to a farm dam/lake or other waterway without having to go through a polluted area first (e.g. through horse holding areas or past manure storage areas). Plan to create an area that has a high level of vegetation for this runoff water to pass through before it enters a waterway, this vegetation will help to filter out any nutrients and pollutants. If you have excess water on a regular basis, then there may be a case for putting in more water storage tanks.

Water can then be reticulated from the storage tanks to where it is needed on the property (see the section *Reticulating water*).

If the water in the tanks runs out and you rely on this as your main water supply, then water will need to be bought. This will need to be delivered by water tanker and it works out relatively cheaper to have a larger tanker deliver the water as most of the cost is associated with the transport. Make sure you plan to have space and adequate surfacing for water trucks to enter and turn around on the property.

Farm dams (ponds)/lakes

Ranging from man-made to natural, these areas can be a real boon on a horse property if managed well. Farm dams are another valuable way of catching and storing rain water that would otherwise run off the land. A farm dam can be sunk into the ground on all sides or be formed by retaining walls of compacted soil (on the side that water exits the farm dam), either way farms dams should preferably be shaded to reduce evaporation. However, trees that will develop large roots should not be planted into or near the farms dams retaining walls. Root growth or death can threaten the integrity of the dam's wall and lead to water leakage over time.

As well as being a useful supply of water for stock, a farm dam can be an attractive recreation area and provide habitat for wildlife. Some of this wildlife (such as insectivorous birds and insectivorous bats) will be beneficial to the property because they eat pest insects in particular. A farm dam can also be stocked with fish, some of which will eat the larvae of mosquitoes; seek qualified advice about what type of fish to stock it with – make sure you are not introducing a pest species. Frogs eat lots of pests too, so encourage them to live in your farm dam by creating inviting areas for them with rocks and suitable fauna.

On a very small horse property, you need to weigh up whether a farm dam and its riparian zone will take up too much space that could otherwise be used for pasture. In this case, it may be better to spend the money on extra water collection tanks.

If you plan to have one or more farm dams, keep in mind that if space is limited, it is generally better to have one large farm dam than several small ones, because small farm dams are not as effective. In a hotter climate, a farm dam needs to be large and deep (at least 4m/13ft) in order to keep the temperature low and reduce evaporation. Small, shallow farm dams stagnate sooner due to increased evaporation rates.

A well planned and managed farm dam is fenced off from stock, so it is not necessary to have one in each paddock. Therefore, one large, strategically placed farm dam will work much better. For example, a farm dam can be placed in a high position on the land from where it can collect rain **and** water from land that is higher still. There is potential here to spare resources by using gravity to feed the water to where it is needed. Again, get qualified advice about the positioning of a farm dam before construction. In areas that have problems with salinity (some parts of Australia and the USA for example), a large body of water in the wrong position can create or increase an existing salinity problem.

These areas can be a real boon on a horse property if managed well.

If the farm dam is to be used for reticulation, then water should be taken from the middle levels. If water is drawn from the top 20cm (8ins) of a farm dam, it is more likely to be contaminated with potentially harmful micro-organisms. However, water taken from near the bottom of a farm dam will be colder and lower in oxygen due to the action of micro-organisms that use oxygen to break down any organic matter at the bottom of it.

If a layer of algae builds up on the surface of a farm dam, you can put a few bales of hay (still in their strings) into it. These bales will float and the algae will stick to them. They can be removed after a few weeks and used as mulch or put onto the manure pile. As well as helping to clear the algae, they also provide a floating platform for wild birds.

Periodically, just before or during a wet period, it is also a good idea to siphon or pump water from the bottom of the farm dam out on to the paddock. This results in esh, oxygenated water replacing the water from the bottom of the farm dam, hich is low in oxygen.

Check with your local authority before having a farm dam constructed as you ay need permission. As already mentioned, expert farm dam construction advice ould also be sought before construction; some soils, such as sandy soils with no ay content, will not hold water. Clay can be brought in, but it adds greatly to the xpense. The farm dam must be compacted properly with heavy machinery when onstructed and the spillway should be placed properly in order to be effective. If a rm dam fails and causes damage to a neighbour's property, you may be liable. erefore, a farm dam should be installed by a qualified and insured earthmover to duce the chance of failing and it must be managed properly if it is not to become stagnant, toxic pool.

sing water

the planning stage, you need to consider how you are going to get water to the rious parts of the property (reticulation) and whether you are going to water rts of the property (irrigation).

If you live next to a waterway and intend to use this for watering your stock, it ay be necessary to pump the water up to holding tanks and distribute it to water ughs from there. This would protect the stream bed and bank from trampling d provide clean, healthy water for your horse/s.

If you intend to use water from the river or stream to irrigate pastures, then you ay need to apply for an irrigation licence (your local Council should be able to sist). If it is necessary to install a water crossing, this should be fenced on both des and the base concreted or in-laid with rocks to avoid further land degrada- n. You may need to seek permission from your local authority in order to dertake any work on a watercourse.

eticulating water

ater can be reticulated around the property from the water source (mains, well, lding tanks etc.). This can be done using polyurethane pipe ('polypipe') that runs troughs, automatic drinkers in stables and surfaced holding yards, and also to rinkler systems if required. If the water is from holding tanks, you will need an ectric or solar powered pump.

If the land is undulating, an extra tank can be positioned in a high area and ater from the tanks near the buildings can be pumped to this (header) tank. The gher up a hill this tank is positioned, the more water pressure there will be. This nk can then gravity feed troughs etc. Using a header tank also creates more ace in the tanks near the buildings so that next time it rains they can fill up again. This system also means that instead of the pump having to start up ever time you turn on a tap, it only has to operate occasionally to pump the water up th hill to the header tank. Another possible option involves pumping water from farm dam/lake to a header tank and then using this water for the toilets, washin horses, watering stock etc., saving the cleaner tank water for drinking and washin in the house (if the house does not have 'mains' water).

If you are relying on an electric pump, it is also a good idea to have a petr pump that can quickly be attached in the case of a fire. This is most important in country that is likely to have a bush fire; when a fire occurs, the power supply i often the first thing to go down, rendering you unable to fight the fire. A portabl petrol pump is useful in that it can also be moved around the property if necessary

If using polypipe, think where it is going to run, usually underground or alon fence lines. If you need to excavate trenches, it is better to only hire the machiner once and it may also be easier to do before any permanent fences are put in. Pla where you might need taps etc. and have connection points put in at this stage You may need to isolate certain areas for repairs etc., so plan where to put sto valves.

The diameter of the water pipe leading into the trough should be of sufficier size to allow refilling of the trough in an acceptable time. Typically 32mm is starting point in a non-mains pressure type reticulation system.

Water troughs

Avoid putting a water trough near a gate, as they creates even more wear and te in this area, or in valleys, as horses will traffic downhill to them, creating th conditions for erosion. Low lying areas tend to be wet already, so a trough place in a dip will tend to create a muddy area.

Also, avoid placing a trough in the corner of a paddock where a horse may be i danger of being trapped by another horse.

Troughs can be placed so that they can be accessed from two paddocks t save costs, unless you have a double fence and then the trough will not reach bot areas.

A trough should be set on a level sand base and surrounded by a firm surfac that will resist hoof action. Place troughs in an area where the ground drains well i wet weather and place gravel around the trough to reduce mud and dust.

Because this area will be bare of grass, also consider placing a trough on th windward side of the stables/surfaced holding yards or other facilities, where it w protect the area against bush fire if there is risk of this occurring.

A reticulated water supply usually relies on a float-valve to control the water-flow into the trough. If the float valve is a type that horses can tamper with, it should be covered to prevent this happening.

A trough can be made from concrete, plastic or steel and their sizes can range from a small plastic or metal 'automatic drinker' usually situated in a stable or surfaced holding yard, to a larger trough for a paddock. A round trough with a one metre diameter or a bath sized rectangular trough is ample for a small group of horses. Keep in mind that some horses will try to splash around in a large trough.

An old bath can be used as a trough on a horse property, particularly if the water supply is not automatic (e.g. just using a hose attached to a tap) or again a float can be fitted if you want reticulated water. Baths are easy to clean and are fine for horses as long as they have no sharp edges that will cause injuries.

Whatever you use to supply the water, it and its access must be checked daily (even more often in hot weather), as it can be knocked over, become contaminated (e.g. by a dead rodent/bird or manure), the horse/s may not be able to get to the water for some reason (e.g. someone accidentally shuts a gate which prevents entry) troughs can break down, streams can stop flowing and water can freeze in cold areas. Horses cannot break ice on troughs and will die of thirst in this situation.

—

This is another advantage of **The Equicentral System** – there is only one water trough to check and it is best placed near the house – see Appendix: *The Equicentral System*.

—

Avoid putting a water trough near a gate.

Irrigation

Irrigation is not essential on a horse property and it is quite possible to manage without it. Installing irrigation can be expensive both in monetary and environmental terms. Changing the pasture species to ones that manage better with less water and improving water retention on the land by improving the water holding capacity of the soil/reducing compaction are more viable options. In fact, you should always be looking to improve the soil.

Water for irrigation comes from farm dams/lakes or ground water via a well, mains water is not usually an option due to the expense.

Irrigation is installed for many reasons including:

- The growing season for pasture can be extended. For example, if rainfall is a limiting factor but there is enough sunshine, irrigation will allow plants to still grow

- Plants that would not usually be able to be grown can be. For example, in a climate that has hot dry summers, plants that are adapted to wet hot summers can also be grown.

- Pastures look better; greener pastures are more aesthetically pleasing. This is after all why people keep 'lawns' green.

- For growing commercial crops e.g. crops or Lucerne hay.

If the property already has an irrigation system in place, your main concerns are whether the system is operating efficiently without water wastage.

There may still be things that you can do to improve the system:

- Make sure the system is not putting more water on the ground than it can cope with. A maximum of 10mm (0.4ins) per application is recommended to avoid fungal problems and waterlogging, however if the soil is compacted any amount of water will run off without soaking in.

- Ensure the plants are of a type that will benefit from irrigation. You need to either sow pasture plants that will benefit from the water, or do not water. There is no point in watering plants that will not grow at that time of year or in watering a paddock full of weeds. Incorrect irrigation can actually bring on more weeds.

- Keep in mind that some plants cannot cope with the higher levels of salt sometimes present in well water.

- Carry out good pasture management (e.g. paddock rotation, manure management etc.) as for non-irrigated pasture. Rotations will need to be more frequent due to the quicker herbage growth.

- Aim to be careful with the water, whether it is from a dam/lake or a well and minimise usage by watering at the right times; do not irrigate during the day in hot weather - early morning or later in the evening is best. Use timers so that drinkers do not get left on.
- Check and maintain the system before a dry period/time of year. Protecting sprinklers from horses and freezing temperatures can be difficult but not impossible.
- Prevent water from irrigators spilling onto roadways or other un-pastured areas. This is wasteful and understandably upsets neighbours.

If the property does not have an irrigation system in place, you need to weigh up the cost of installation with the costs of sowing different pasture and the long term benefits to you of both options. It may be that it is not worth the extra expense taking into account the chances of failure of the system e.g. if the well stops producing water, compared to the costs of using surfaced holding yards etc. for the horses and buying in extra feed.

Pasture can usually be vastly improved without irrigation. Even when the shortfall in feed has to be made up by purchasing feed, it may still be far lower than the cost of irrigation (including the environmental and labour cost). You should also employ strategies to ensure that any water arriving on the property (e.g. rainfall) remains on the property for as long as possible, especially if you live in an area that has long dry periods. Improving the water holding capacity of your land is something that should always be done before irrigation is installed and this may be enough in many cases.

If you plan to install a new system, aim for ease of use and economy. In small paddocks, the system can be positioned around the fence line. A portable system using hoses and free standing sprinklers can be used for the middle sections if necessary. In larger paddocks, it may be necessary to position permanent sprinklers in the middle of the paddock, as well as around the edges. Popup sprinklers work well with horses as they can sink underground out of harm's way when the paddock is in use. Post fastened sprinklers are protected from horses if they are currently grazing the paddock. If placed high above the ground (e.g. 1.8m / 6ft), they seem to get less damage from horses than if lower. Where possible, these should not cause obstructions to machinery for paddock maintenance.

If you are thinking about installing irrigation, it is a good idea to talk to people who already have it to get some ideas. Some irrigation suppliers give a design service, however this will be aimed at using all of their products whereas it may be more cost effective to use products from various outlets. Make sure the design is not more elaborate than you need, but if you plan to expand now is a time to consider your needs of the future.

Planning for clean water

It is a good idea to work out a water management plan before you fence a property

These are some of the points to keep in mind:

- Do not use steep hillsides to paddock horses, as the soil and manure is washed more readily into the water system.
- Horses on steep land also cause erosion which leads to more soil ending up in the waterway.
- Water run-off from slopes should be directed *across* the landscape rather than directly down any hills when possible. This slows the water down, reduces the damage that it can do and gives it more opportunity to soak into the ground.
- This process can be helped by fencing paddocks and laneways along contour lines whenever possible and keeping any fences that go downhill as short as possible.
- Any streams and rivers that go through or border the property (and any farm dams), should be fenced off and if necessary the water can be reticulated around the property.
- Don't allow bare areas to form anywhere on the property, bare soil will be blown or washed away and is usually deposited in the waterway. They are also an invitation for weeds.
- Take care with fertilisers, either chemical *or* organic. Even though a large amount may be recommended, it may be better to apply it in stages, so that any run off will be reduced. This is a 'catch 22 situation', as fertiliser, correctly applied, will help plants to grow vigorously, which in turn will vastly reduce the flow of pollutants into the waterway by trapping and using the nutrients. Correct fertilisation is important, as is the timely application of that fertiliser. Remember it is usually possible to improve your soil with minimal use of fertiliser; ensure that you are only using what is needed.
- Install rain gutters and either collect water in tanks or divert the run off so that it does not run across holding yards and manure storage areas.
- Site any new buildings and holding yards as far from a waterway as possible, aim for at least 100m, again you will need to check with your local authority, and on the highest parts of the property. Make sure that contaminants do not leach

into the soil by siting structures on an impervious layer such as concrete or compressed limestone.

- Use environmentally friendly biodegradable products for washing horses and washing horse gear; some shampoos actually prevent water from entering the soil.

Use environmentally friendly biodegradable products for washing horses This horsewash contains footings of gravel which help to filter out any impurities after washing and before the water is used to irrigate the pasture.

—

The subject of horses and water is a very important subject with regards a horse property so make sure you read the information in *The Equicentral System Series Book 2 - Healthy Land, Healthy Pasture, Healthy Horses*.

—

Vegetation planning

The following information is an excerpt from the second book in this series *Th Equicentral System Series Book 2 - Healthy Land, Healthy Pasture, Health Horses* where there is more information about horses and vegetation including **the benefits of trees and bushes, trees and bushes as habitat for wildlif fodder trees and bushes, buying and planting, protecting vegetation fro horses, poisonous trees and plants**, and **information about the dangers c allowing horses access to lawn mower clippings**.

Windbreaks and firebreaks

Trees and bushes as windbreaks help to stop moisture loss in soil (by approx mately 20% – 30%). They reduce evaporation from the soil, enabling the plants utilize the moisture more effectively and increasing growth periods and therefo yield. They also reduce erosion by wind on dry soils.

They need to be across the line of the prevailing winds to be effective; plan plant shelterbelts that will benefit areas such as surfaced holding yards ar paddocks as well as a house and other buildings. Double fences and laneway between paddocks provide an excellent area to plant windbreaks.

A good windbreak slows the wind down, protecting everything on the lee sid Windbreaks should be constructed so that they *slow* the wind speed by 60% to k effective. Solid windbreaks cause the wind to gust over them and crea turbulence on the lee side. Permeable windbreaks are better because they simp slow the wind as it goes through. A good windbreak will extend its effect 20 time its height downwind and 5 times its height upwind of the windbreak. For shelterbelt of trees to do its job correctly, it should be at least 20-25 times as lor as it is high. The trees need to be stepped by planting in at least three rows wi the tallest trees in the middle. This is achieved by planting different species th have different heights at maturity.

If fire is a risk in your area/country, windbreaks should also be designed wi this in mind. You need to know what direction fire is likely to come from and pla at right angles to this direction.

They should not be positioned too close to buildings or surfaced holding yard In particularly high risk areas where the windbreak is too close to building remove the middle story of plants to reduce the risk of bushes igniting the tree canop

Even fire resistant trees and bushes will burn eventually, however they do off some protection. They reduce the wind speed and this reduces the speed ar intensity of the fire. Trees absorb the radiant heat, protecting animals and buildinc on the lee side. They also catch airborne burning material, stopping it from flyir into other areas and starting spot fires. Organic matter around the base of these trees and bushes needs to be kept clear as this counts as fuel.

Aim to plant flame resistant varieties of trees if fire is a risk in your locality; flame resistant trees and bushes tend to have certain characteristics in common. These characteristics are:

- They are usually slow growing (e.g. deciduous trees).
- They usually have a high salt content (such as mangroves and salt bushes).
- They should not collect litter within the canopy or on the bark (smooth barks rather than stringy or paper barks).
- They usually contain high moisture levels in their leaves (rainforest plants, cacti, succulents and most fruit trees).
- They usually contain low amounts of volatile oils.

Always seek advice from your local fire authority regarding any use of tress/bushes with regard to fire prevention.

Every area differs in what can be grown and in what is available. A local land/soil conservation group will usually be happy to provide you with the names of useful shelterbelt and firebreak trees for your district.

A good windbreak slows the wind down, protecting everything on the lee side.

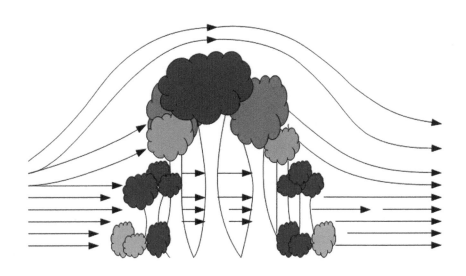

Revegetation of steeper land

In many cases, land is sold for horsekeeping that is actually too steep for grazing of large animals. Steep land is far harder to manage well than is the case with flatter land.

Steep land can be problematic for the following reasons:

- Weed management is more difficult because it is not safe to drive machinery (such as a mower) on steep slopes. Manure is also more difficult to manage for the same reason, because it is not safe to drag a pasture harrow on steep slopes.
- Manure is more readily washed into the waterways on steep land.
- Gravity and water (rain) result in soil being washed downhill. This process is exacerbated if the land is overgrazed and therefore there is not enough vegetation to hold the soil together.
- Large grazing animals add to this problem (in addition to overgrazing if this is allowed) by physically 'pushing' the soil downhill. This speeds up the process of soil erosion and soil being washed into the waterways.

In addition, horses should not be permanently kept on steep land as this will lead to joint problems over time. Horses are fine grazing up and down hills of course, but they need to be able to get to flat areas to 'loaf' (sleep and 'hang around' together). This is because they are not comfortable/cannot sleep either standing in that characteristic position with one leg resting or lying flat out, on a hill side. Horses have virtually no sideways movement in the joints of the legs, so forcing them to stand on the side of hill puts an unnatural force on these joints which over time will lead to joint problems.

Sometimes it is easier, and much better for the environment *and* your animals, to fence off any steep areas on the property, re-vegetate these areas with trees and bushes, and concentrate on growing pasture on the flatter areas of the property.

Easy areas to increase vegetation

A good figure to aim for is for approximately 30% of a given area of land to be vegetated with trees and bushes. There are various places on a horse property where trees and bushes can be planted where they will not necessarily detract from the available grazing (and may enhance it), and they will be able to carry out their many beneficial functions. As mentioned previously, well positioned trees and bushes may actually increase the yield from your property.

You should research which trees and bushes are best in your area in terms of fire resistance and check that they are not are not designated weeds or classified s being poisonous to horses. Whenever possible, plant trees and bushes that are ative to the area as these will be best adapted to local conditions, soil etc.

Trees and bushes on hillsides and the tops of hills help with fertilisation of land below them by attracting a huge variety of insects and animals (and their dung). The leaves also aid fertilisation and the plant roots hold soil together and therefore reduce erosion.

he corners of paddocks

he corners of paddocks are dangerous if horses are kept in herds, because they an run into a corner while playing and get trapped. In terms of land management, ey are difficult manage because harrowing them (spreading manure) is difficult. or example, when driving around a paddock with pasture harrows you have to top, get off, and kick the manure around because the harrow cannot get into the orners. Corners often end up bare, compacted and full of manure and are erefore less productive than other areas of a paddock.

A good solution is to fence the corners off with a simple electric tape/braid - or a ore permanent fence. Then, a few trees and bushes can be planted in the orner. The situation is now turned around completely; previously unproductive nd potentially dangerous corners will become wildlife havens, providing all the enefits, *and* provide extra shade for your horses etc. – for the cost of a bit of lectric tape or braid and a few young plants.

The perimeter of a property

There should be a double fence around the perimeter of a horse property fo various reasons:

- A double fence means that you have two fences rather than one keeping you horses on the property.

- A 'living fence' of trees and bushes, backed up by more 'permanent' fencin materials such as a simple plain wire, presents a good solid visual barrier t horses moving at speed and is therefore safer. In fact, whereas horses wi attempt to jump an ordinary fence, they will never attempt to jump a 'living fence if it is high enough.

- A double fence prevents your horses from playing/socialising with a neighbour' horses over a fence. As we have seen before, horses should never be allowe to interact over a fence; beside the physical dangers (fence injuries are high o the list of the cause of death or permanent injury to horses), there are th biosecurity issues when horses on different properties interact.

- If you live in or near suburbia it makes it more difficult for well-meaning peopl to feed your horses over the fence. This includes people dropping lawn mowe clippings and other garden waste in with your horses.

- A double fence means that you can turn a potentially unsafe fence into a muc safer fence. A double fence does not necessarily mean double the expense. Fc example if the property has an undesirable fence already in place around th perimeter, it will probably cost less to plant trees and bushes on the inside c that and then place a simple electric fence on the inside of that, than to re-fenc with a new solid fence (which will still only be a single fence).

A row of trees planted around the perimeter of you property provides many advantages.

Between paddocks

Trees and bushes can be planted between paddocks within the property. If you have more than one herd for example, this double fence will allow you to use adjacent paddocks.

As an addition to a shelter

Trees and bushes can be planted so that they enhance man-made shelters or, in some cases, provide total shade for surfaced holding yards. They need to be planted outside the surfaced holding yard so that they are protected from horses by the surfaced holding yard fence. A solid roof can be enhanced with the addition of trees and bushes which will slow the wind without stopping it entirely.

In laneways/driveways

Trees and bushes can be planted in any laneways - if you have to have them in order for horses to take themselves to paddocks (see Appendix: *The Equicentral System*) in which case they may need the protection of a double fence. This second fence can be a simple electric fence. If horses are being led to paddocks, then a second fence will not usually be necessary. The driveway of a property is another example of a good place to plant trees and bushes.

Trees and bushes can be planted between paddocks within the property.

Along contour lines

Trees planted in the correct areas can help to move water in a more desirable direction. For example, by planting a row of trees across a hillside and banking the soil a little, water that previously ran straight down the hill can now be channelled sideways and slowed down. Some of the water can be allowed to pass through the line of trees if necessary, which means that instead of the water all travelling down the same path, it is now spread out and benefits a much larger area, in addition to reducing the damage that water travelling in a single channel can cause.

In wet areas of land

Water-loving trees can help to soak up excess water, which often occurs in areas that were originally very wet areas e.g. wetlands. By planting the correct trees, you can redress this issue without going to the trouble and expense of installing drainage. Aim for local species whenever possible as these will be better adapted to your conditions.

Trees and bushes can be planted around waterways, high ground, eroded areas, in fact anywhere unsuitable for grazing.

Around waterways

Any waterway that is on or borders a property, whether natural or manmade (such as a farm dam), should have a riparian zone around it.

On steep sections within paddocks

Steeper land is difficult to manage as pasture and therefore is ideal for growing larger vegetation instead, see the section **Revegetation of steeper land**.

—

The subject of horses and vegetation is a very important subject with regards a horse property so make sure you read the information in **The Equicentral System Series Book 2 - Healthy Land, Healthy Pasture, Healthy Horses**.

—

The following section/s are directly reproduced from *The Equicentral System Series Book 1 – Horse Ownership Responsible Sustainable Ethical* in order to aid your understanding in this book.

Appendix: The Equicentral System

The Equicentral System is a *total* horse and land management system that we have developed and have been teaching to horse owners around the world for many years now. It uses the natural and domestic behaviour of horses, combined with good land management practices, to create a healthy and sustainable environment for your horses, the land that they live on *and* the wider environment.

There are many examples in various countries including Australia, New Zealand, the UK, the USA and even Panama! This this list just keeps growing as people realise the huge benefits of using **The Equicentral System** in order to manage their horses *and* their land in a sustainable way.

How The Equicentral System works

The Equicentral System utilises the natural grazing and domestic paddock behaviour of horses in order to benefit the land that they live on *and* the wider environment. In turn this benefits the horses and it also benefits you (and your family) because it saves you money and time/labour.

- The main facilities – water, shade/shelter, hay and any supplementary feed are positioned in a surfaced holding yard so that the horses *can always* get back to them from the pasture they are currently grazing.

- The watering points are *only situated in the surfaced holding yard*, instead of there being one in each paddock. If you already have water troughs' situated in paddocks these can be turned off when horses are using the paddock, and turned back on again if and when other animals, such as cows or sheep, are grazing there.

- Individual water troughs/drinkers (or buckets) of course are also needed in any individual yards/stables.

- If possible all of the paddocks *are linked* to this surfaced holding yard area, although only one paddock is in use at any time.

- The gate to the paddock that is currently in use *is always* open, so that the horses *can always* get themselves back to the water/shade/shelter etc. *In short, the horses are never shut out of the surfaced holding yard*.

- Occasionally they may be fastened in the surfaced holding yard (with hay), but this is usually for the purpose of preventing damage to the land and increasing healthy pasture production.

- Apart from trees or bushes that are situated in/around paddocks, **the only shade/shelter is in the surfaced holding yard**. This shade/shelter is very important. It should be large enough for the whole herd to benefit from it at the same time.

The Equicentral System: *all of the paddocks lead back to the surfaced holding yard. There is shade/shelter and water in this central area. Hay can also be fed here.*

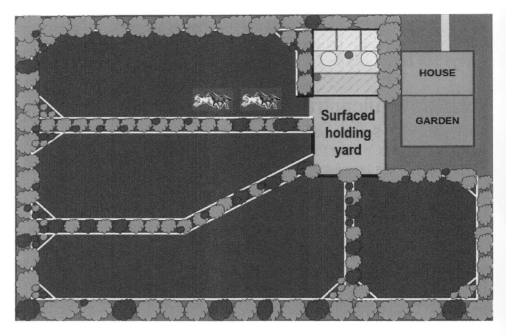

Additional information

- The surfaced holding yard area *can* also be a riding/training surface if that is what you wish. You may prefer to keep it separate or indeed you may not need a riding/training surface - but if you do then this means that the expense of creating this surfaced area has double benefits. This could mean that you are able to afford and justify this surfaced area sooner because you are going to get more use out of it. Also remember, the smaller the property, the more the facilities need to be dual purpose whenever possible so that you have as much land in use as pasture as possible.

- Careful consideration of the surface is required, especially if it is to be a riding/training surface as well. Remember - bare dirt is not an option. Wet mud is dangerously slippery and can harbour viruses and bacteria which can affect the horses legs. Mud will become and dusty when dry, meaning that the horses will be breathing in potential contaminants *and* you will be losing top soil.

The surfaced holding yard area can also be a riding/training surface if that is what you wish. If you do then this means that the expense of creating this surfaced area has double benefits.

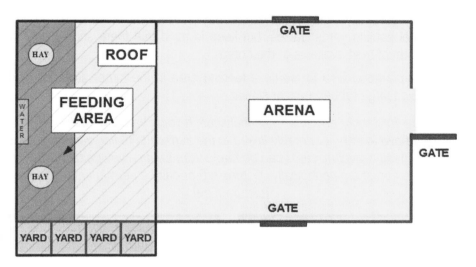

—

See the third book in this series for more information about using a surfaced holding yard as a riding surface as well, *The Equicentral System Series Book 3 – Horse Property Planning and Development.*

—

It is useful if there are also some individual holding yards (or stables if you already have them) – preferably linked to the surfaced holding yard for ease of use. You can then separate horses into them for any individual attention that they may require (such as grooming, supplementary feeding etc.) or for tacking up etc. You can also put the surplus horses in them if you are riding/training one of the herd members on the larger surfaced holding yard.

This system is *not* about food restriction – quite the opposite. It is about transitioning horses to an ad-lib feeding regime of low energy pasture plants and hay (see the section *Changing your horse/s to 'ad-lib' feeding*).so that they no longer gorge and put weight on because of it.

Your pasture may need to be transitioned to lower energy plants. This does not always mean reseeding.

—

This is too large a subject to cover here and is covered in the second book in thi series *The Equicentral System Series Book 2 - Healthy Land, Healthy Pasture, Healthy Horses*.

—

- Hay can be fed in the larger surfaced holding yard if the horses get on w enough; generally horses will share hay. Otherwise, it can be fed in tl individual holding yards/stables, but keep in mind that there should *always* l some form of feed available to the horses.

- It can be a good idea to create a feeding area in the larger surfaced holdi yard using large rubber mats or similar.

  It is useful if there are also some individual holding yards (or stables if you already have them) – preferably linked to the surfaced holding yard for ease o use. These individual yards can be made from swing away partitions if you only require them periodically (picture left) or can be permanently in place (picture right).

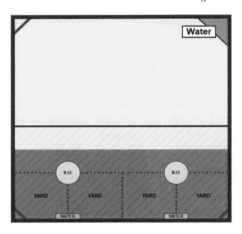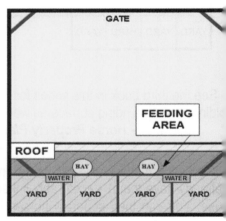

- **The Equicentral System** works best on a property where the horses li together as one herd, otherwise you will need to replicate it for each group horses that you have. However, many of our clients however have done just tl in the case of larger properties with various classes of horses (for examp studs, livery yards etc.).

- **The Equicentral System** assumes that you already have good grazi management in place (rotational grazing) or that you plan to implement Remember, rotational grazing involves moving the animals around the land as herd, one paddock at a time, rather than allowing them access to the whole property at once (set-stocking).

The Equicentral System in practice

This is an example of how **The Equicentral System** works in practice. In this example, the horses are being kept in the surfaced holding yard at night (or in individual holding yards/stables) and out at pasture through the day, but remember - if there is enough pasture, then the horses do not need to be confined overnight unless you have other reasons for doing so.

- In the morning you open the surfaced holding yard/riding arena gate and the horses **walk themselves** to the paddock that is currently in use for a grazing bout (which lasts between 1.5 to 3 hours), the gate to this paddock should already be open (the other paddocks should have closed gates as they are being rested).

- At all times the horses are free to return to the surfaced holding yard for a drink, but they usually won't bother until they have finished their grazing bout.

- After drinking, the shade and inviting surface in the surfaced holding yard encourages the horses to rest (loaf) in this area before returning to the paddock for another grazing bout later in the day.

After a grazing bout, the horses return to the surfaced holding yard for a drink.

- Leaving hay in the surfaced holding yard can encourage even more time being spent (voluntarily) in the surfaced holding yard and less time spent in the paddock.

- At the end of the day, the horses return from the paddock to the surfaced holding yard to await you and any supplementary feed that they may be receiving.

- You simply close the gate preventing them from returning to the paddock for the night, or, if conditions allow it, the horses can come and go through the night as well as through the day.

After drinking, the shade and inviting surface in the surfaced holding yard encourages the horses to rest (loaf) in this area before returning to the paddock for another grazing bout later in the day.

The Equicentral System benefits

The Equicentral System utilises the natural and domestic behaviour of horses to better manage the land that they live on. This system of management has many, many benefits including:

Horse health/welfare benefits:

- It encourages horses to move more and movement is good; a grazing horse is a moving horse. A recent (Australian) study showed that horses in a 0.8 Ha. paddock walked approximately 4.5km per day, even when water was situated in the paddock; additional movement to the water in the surfaced holding yard, therefore, further increases this figure. More movement also means better hoof quality as the hooves rely on movement to function properly. Remember a healthy, biodiverse pasture encourages more movement.

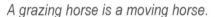

A grazing horse is a moving horse.

- It *maximises* time spent grazing for horses and aims to avoid food restriction. By confining horses initially and when the weather is very dry or very wet (or in order to transition horses that have been on restricted diets in the past), so that the pasture begins to improve, becomes more biodiverse etc. they will be able to graze more in the future because healthy pasture can withstand more grazing.

- Horses are not being forced to stand in mud, especially around gateways when weather conditions are wet. Horses are not good at coping with continuous wet conditions – hence the ease with which they develop skin conditions such as greasy heel/mud fever. Remember - in the naturally-living situation, they will *take themselves* to higher, dryer ground to loaf, even if the areas that they graze are wet. When horses are fastened in wet paddocks, they do not have this choice.

The horses will be able to graze more in the future because healthy pasture can withstand more grazing.

- Eliminating mud also means eliminating dust because they derive from the same thing (bare soil) at either end of a spectrum. Apart from the obvious benefits to not losing your top soil, this means that horses (and humans) do not have to cope with living in a dusty environment.
- Horses move around a paddock in a natural fashion, choosing what to eat. As rotational grazing increases the diversity of plants in a pasture, the horses benefit from access to a larger variety of plants. This means that the horses eat a more natural, varied diet. In addition, healthier plants are safer to graze than stressed, overgrazed plants (more fibre and less sugar per mouthful).

—

See the second book in this series *The Equicentral System Series Book 2 - Healthy Land, Healthy Pasture, Healthy Horses* for much more information on this subject.

—

The stress of not being able to get to food at will, along with all of its associated problems, is removed. Horses with different dietary requirements can be catered for with the addition of supplementary daily feed (in a separate but preferably adjoining area).

The surfaced holding yard is seen by the horses as a good place to be and therefore, if and when it is necessary to fasten them in there, (bad weather, for the vet etc.) they are not stressed.

The horses now have choice; they can choose as a herd whether to graze, walk to the 'water hole', snooze in the shade etc. Instead of us deciding when a grazing bout will start/finish, the horses can decide for themselves. These are all behaviours that naturally-living horses take for granted, but domestic horses are usually 'micro managed' in such a way that a human decides where they will be at any point in the day. This might not seem like a big deal but it really is.

The Equicentral System provides a 'home-range' whereby the horses can access the available resources in a more natural fashion.

ime saving benefits:

You do not have to lead horses in and out to the pasture. The horses are waiting for you close to the house, or at least in an area where you need them to be, much of the time. If they are currently grazing, you simply call them; horses soon learn to come to a call for a reward. This means that when you return from work and the weather is bad, you do not need to trail out in the wind and the rain to bring them in, they will be waiting for you in the surfaced holding yard.

You do not have to spend time carting feed around the property (or keep a vehicle especially for the job) because the horses *bring themselves* to the surfaced holding yard for feed. The horses *move themselves* around the property, *taking themselves* out to the paddock that they are currently grazing, and bringing *themselves* back for water and feed.

The single water trough in the surfaced holding yard is all that you have to check each morning and night, saving you having to go out to a paddock and check the water.

It is far quicker to pick up manure from the surfaced holding yard than from pasture if you collect your manure.

Any time you save can be spent on other horse pursuits such as exercising them!

You do not have to spend time carting feed around the property.

Cost saving benefits:

- Money spent on the surfaced holding yard is money well spent, as this area
  used *every day of the year* for *at least twelve hours* a day, even if you are r
  also using this surface for riding/training.

- Money spent on vet bills for treating skin and hoof conditions is reduced
  totally eliminated.

- The expense of installing and/or maintaining a water trough in each paddock
  spared.

- The expense of installing individual shade/shelters in each paddock is spare
  Instead, one large shade/shelter is erected at the side of or over the surfac
  holding yard, which means you may end up with a partially covered all weath
  riding/training surface if you are multi-tasking this area!

- This large shade/shelter will be in use *every day* of the year, unli
  shade/shelters that are situated in paddocks and are only in use when t
  paddock is in use. Remember - if you are rotating your paddocks (as part of
  rotational grazing management system), then this means that each paddock w
  be empty, and any shade/shelters situated in them will be unused, for a lar
  part of each year.

- Annual maintenance including time and expense, of numerous shelte
  (especially if they are made of wood) is avoided adding to the cost effectivene
  of **The Equicentral System**.

- Many horse properties already have the facilities required to implement **T
  Equicentral System**. Often the required infrastructure either already exists or
  horse property, or the property needs minimal changes.

- Laneways (and their associated costs) can be kept to a minimum. In areas that *do* require laneways, any money spent on surfacing them is well utilised as the laneways will be used by the horses several times a day.

This large shade/shelter will be in use every day of the year, unlike shade/shelters that are situated in paddocks and are only in use when the paddock is in use.

- Better land management means more pasture to use for grazing (and safer healthier pasture) and more opportunities for conserving pasture (as 'standing hay' for example) or making hay. This all leads to much less money being spent on bought-in feed.

- You do not need to buy and maintain a vehicle for 'feeding out'. The horses come to where the feed is stored rather than you having to trail around the paddocks 'feeding out'.

- Setting up **The Equicentral System** will not devalue your land. It will actually increase the value of it through good land management. Likewise, if you sell the property, the next owner can choose to set up a more traditional management system by putting water and shade/shelter in every paddock if they wish.

Safety benefits:

- Horses move themselves around the property, therefore there is less unnecessary contact between humans and horses. This is an important point if you have (usually less experienced) family or friends taking care of your horses when you are away. **The Equicentral System** allows them to see to your horses without them having to catch and lead them around the property.

- It reduces or eliminates the incidence of horses and people being together in a paddock gateway. When horses are led out to a paddock, they can be excited because they are about to be freed; and when they are waiting at a gate to come back in for supplementary feed, they are keen get through the gateway in the other direction for their feed. Horses can crowd each other and human handlers can become trapped. These situations are very high risk on a horse property.

- Depending on its position, the surfaced holding yard can be a firebreak (for your home) and a relatively safe refuge in times of fire/storm/flood for your horses. The layout of the property may result in the horses being pushed (by rising water) back towards the surfaced holding yard in a flood. Assuming the surfaced holding yard is built on higher ground, this can save lives! By training the horses to always come back on a call, you can get them into the surfaced holding yard quickly in any emergency situation. This makes it far easier for you, your neighbours, or the emergency services to evacuate your horses if necessary in emergency situations.

Land/environmental management benefits:

- **The Equicentral System** is a *sustainable* system that acknowledges that a horse is *part of* an ecosystem, not separate to it.

- **The Equicentral System** complements a rotational grazing land management system and allows for the fine tuning of it. Remember - rotational grazing encourages healthy pasture growth and aids biodiversity by moving the animals to the next grazing area before they overgraze some of the less persistent plant varieties.

- With good land management, the productivity of biodiverse, safer pasture should *increase* rather than decrease over time, leading to fewer periods when it is necessary to fasten horses in the surfaced holding yard over time. Remember - biodiversity is good for horses *and* good for the environment.

- It *vastly* reduces land degradation that would be caused by unnecessary grazing pressure. The horses *voluntarily* reduce their time spent on the pasture

They will tend to spend the same amount of time grazing (as they would if they were fastened in a paddock for 24 hours), but will tend to carry out any other behaviours in the surfaced holding yard.

They will tend to carry out any other behaviours in the surfaced holding yard.

- They prefer the surfaced holding yard not least because, if it is situated near the house, or at least in an area that they can see you coming towards them, no self-respecting horse will miss an opportunity to keep watch for the possibility of supplementary feed! The water and shade in the surfaced holding yard also encourages the horses to loaf in this area. If the horses are allowed to come and go night and day they will reduce the grazing pressure (grazing pressure being a combination of actually eating but also standing around on the land) by approximately 50%. If you fasten them in the surfaced holding yard (with hay) overnight, you will further reduce the grazing pressure by about another 50% (making a total of about 75%). This reduction in grazing pressure will make a *huge* difference to the land.

- The corresponding compacted soil/muddy areas that surround water troughs and paddock shelters, as well as the tracks that develop in a paddock are avoided. Bare/muddy/dusty gateways are also a thing of the past as horses are *never* fastened in a paddock waiting to come in. Don't forget that the idea is to reduce any unnecessary pressure on your valuable pasture and increase movement. Remember - the reason horses stand in gateways is because that is usually the nearest point to supplementary food; they are either fed in that area or their owner leads them from there to a surfaced holding yard or stable to feed them. If the gate is closed, they stand there; if the gate is open, they bring themselves into the surfaced area, which becomes their favourite loafing area.

—

Managing your land in this way results in less or no soil loss, in fact if you manage your land well you should be able to increase soil production – see the second book in this series *The Equicentral System Series Book 2 - Healthy Land, Healthy Pasture, Healthy Horses*.

—

- It reduces the area of land used for laneways and therefore the land degradation caused by them – by minimising laneways as much as possible. This is done by creating a layout whereby the paddocks lead directly to the surfaced holding area or by creating temporary laneways. If paddocks are already fenced and laneways are in place then this system utilises them efficiently and safely e.g. the horses are not fastened in narrow areas, they can spread out when they reach the paddock at one end or the surfaced holding yard at the other.

- It increases water quality – by minimising or eliminating soil and nutrient runoff. Rotational grazing maintains better plant cover – the absolute best way to keep soil and nutrients on the land and out of the waterways.

- Strip grazing is usually easier to set up because the water point is back in the surfaced holding yard, meaning that the fence only has to funnel the horses back to the gate, without having to take the water trough position into consideration.

- Hay is fed in the surfaced holding yard area rather than the paddocks allowing for better weed control.

Public perception benefits:

- **The Equicentral System** helps to create a positive image of horsekeeping.

- **The Equicentral System** is most likely to be regarded as a good way to manage land by landowners, the general public and the local authorities. There is a general expectation that land should be well managed - e.g. less mud/dust and fewer weeds rather than more mud/dust and weeds.

- **The Equicentral System** fulfils this expectation, creating a positive image of horse ownership rather than a negative one. This is an important point remember - in some areas legislation is being pushed forward to reduce horse keeping activities due to the negative image caused by the often poor land management practices on many horse properties.

- In particular, as horses are often kept on land that is leased rather than land that is owned by the horse owner, the landowner usually, and quite rightly, expects to see good land management taking place. Of course horse owners that own their own land should, and usually do, want the same.

There is a general expectation that land should be well managed - e.g. less mud/dust and fewer weeds.

Manure and parasitic worm management benefits:

- The manure, along with the horses, comes to you. More manure is dropped in the surfaced holding yard and much less in the paddocks (as much as 75% less if you fasten the horses in the surfaced holding yard/s at night with hay). This allows for much better manure management.

- If you usually collect manure that is dropped on pasture then it is physically easier to pick up manure from the surfaced holding yard/s.

This collected manure can then be composted (which also reduces parasites on your property, as thorough composting can kill parasitic worm eggs and larvae).

Composted manure is much better 'product' than 'fresh' manure.

- Less manure on the pasture is less importunity for parasitic worm larvae to attach to pasture plants.

Better manure management also means less reliance on worming chemicals.

- The extra pasture created by managing the land better increases the possibility of being able to 'cross-graze' (graze other species of animals on the land). This further reduces parasitic worms on the land in the most natural way possible, because parasitic worms are what is termed 'host specific', meaning that they can only survive when picked up by the host animal that they evolved alongside.

- Rotational grazing also aids in parasitic worm management by increasing the time that a given area of pasture is resting, which means that *some* of the parasitic worm larvae (on the pasture) dry out and die as they wait - in vain! - for a horse to eat the plant that they are attached to.

—

See the second book in this series ***The Equicentral System Series Book 2 - Healthy Land, Healthy Pasture, Healthy Horses*** for detailed information about manure management including many novel ideas such as how chickens can be used to help you to manage horse manure.

—

Manure dropped on the surfaced yard rather than pasture is also far preferable in terms of parasitic worm management (no plants for any larvae that hatch out to attach to).

Implementing The Equicentral System

This section describes some of the practicalities of implementing **The Equicentral System**. In some sections you will be referred on to one of the other books in the series because there is too much information to be covered here.

On your own land

Obviously, this is what most people aspire to; having their own land. If you are in this fortunate position, then you are free to set up **The Equicentral System** and reap the benefits.

If you are fortunate enough to own your own land then you are free to set up **The Equicentral System** *and reap the benefits.*

On small areas of land

The Equicentral System is the ideal way to manage horses when there is only a small area of land available for grazing. In fact, it is the only way that will ensure that the horses have grazing available to them and at the same time, land degradation is not created. Of course it will mean that the horses are not able to graze as much as they or you would like, but at least the grazing they do have will be 'quality' grazing rather than standing around on bare, dusty/muddy, weedy land. So don't ever think that your situation would not support **The Equicentral System**, because it will.

At least the grazing they do have will be 'quality' grazing.

Source - Alayne Blickle of Horses for Clean Water - USA.

A common scenario, especially when people lease land, is that they have just on paddock (picture A). Even in this situation it is not difficult to set up **Th Equicentral System**. You can still make huge changes to your management of the land and horses by creating a fenced hard standing area by the gate (surface holding yard) preferable with a shade/shelter. The 'paddocks' can fan out from thi area (picture B). The facilities/fencing can be made temporary/relocatabl materials including sectional holding yard fences and rubber paving mesh for th surfaced holding yard and electric fencing for the internal fences.

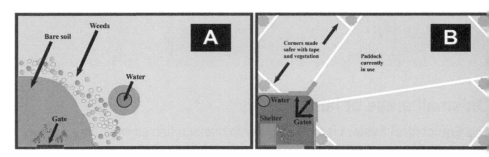

On large areas of land

The Equicentral System works well on a large 'mixed use' property as well as on a large horse property such as a stud. Other species of grazing animals respond well to having a centralised area for resources so it is possible to set up multiple central holding areas on large properties that have different types of grazing animals. Likewise, a large horse stud that has various age groups of horses can also have multiple central points.

In this example (on a 100 ac/40 he property) there is an **Equicentral System** *set up for the horses, positioned near the house as the owners will be handling the horses much more regularly than they will the cattle. The central area for the cattle is at the bottom of the hill, well away from the house. Occasionally the owners can bring the cattle up the hill (via the driveway or through various paddocks) so that the horse paddocks can benefit from cross grazing.*

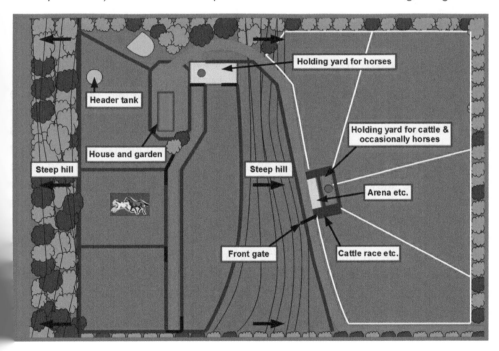

*Two **Equicentral Systems** in place. When the youngstock need to be brought up to the main yard near the house, the mares and foals are fastened in one of the paddocks.*

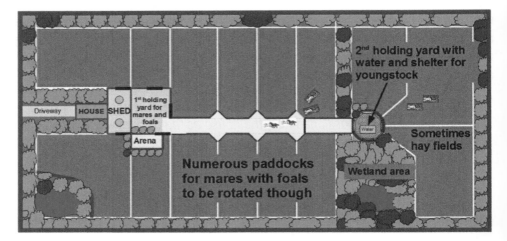

In different climates

The Equicentral System works equally well in any climate, whether it be temperate/wet/tropical/dry/arctic etc. This is because you are providing a 'home range' for the horses and allowing them (most of the time) to make the decisions about when to graze and when to shade/shelter. So, when insects are particularly problematic, they can *take themselves* to the shade to escape them and when they need fibre, they can decide for themselves when to go out to graze or to stay in the shade/shelter and eat hay (if you leave 'ad-lib' hay in the shelter). When it is very cold and wet, they can decide to shelter mainly at night and graze mainly by day. You only need to step in and 'micro manage' them when they are about to put too much pressure on the land that would lead to less pasture in the future.

Yes it is good to understand different land class types and understand what sor of soil you have, but initially it is more important that you understand that land and climates tend to range from too wet to too dry.

Either way, as long as you have an area to allow the horses to remove the pressure voluntarily, as well as involuntarily when you decide that the land needs a hand; you will see a reduction in, and eventually an elimination of dust and mud and its associated problems.

Using existing facilities

If your land already has facilities in place, **The Equicentral System** can usually be implemented without making any major structural changes to your property. **For example:**

- Hard standing that is already in place around any farm buildings/stables etc. should be able to be utilised as a surfaced holding yard. So, if you already have a 'stable yard' that has hard-standing with the paddocks leading out from this area, then you already have a great set up.

- In many cases it is just a matter of leaving the gate to the paddock that is currently in-use open so that the horses can get back to this area, rather than fastening them on the other side of the gate.

- Old farm buildings can usually be used to great effect, as long as they are safe and have a high enough roof for horses. Such buildings often already have hard standing in and around them.

An old farm building such as this would be great for converting to a 'run-in shed' for horses.

By implementing rotational grazing and always having the gate open to the paddock they are currently using, the horses will bring themselves back to the yard and stand on the surfaced area, rather than stand in the gateway and create mud.

- You can turn off the water in the paddocks (or stop carrying water out to the paddocks!), and set up a water trough on the hard-standing area.

- You may want to create extra shade/shelter for the horses by using the existing buildings to fasten 'shade sails' from, or extend the roof area with a more solid style of roof.

- If you have a block of stables you may decide to open the fronts of some or all of the individual stables boxes to create a 'run-in shed'. Keep in mind that, for various reasons such as tacking up, health care/vet work/trimming/shoeing, supplementary feeding etc., it is still useful to have some individual holding areas.

- Surplus stables can be used for storing hay etc.

By implementing rotational grazing and always having the gate open to the paddock they are currently using, the horses will bring themselves back to the stable yard and stand on the surfaced area, rather than stand in the gateway and create mud.

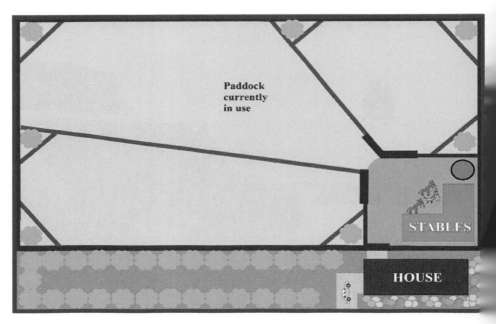

You may want to create extra shade/shelter for the horses by using the existing buildings to fasten 'shade sails' from, or extend the roof area with a more solid style of roof.

Source - Alayne Blickle of Horses for Clean Water - USA.

On land that you lease

A landowner should be happy for you to implement a system of management that is going improve their land value. For example, horse owners often lease land from farmers and most farmers understand the value of a rotational grazing system.

If you need more land and you are already demonstrating good land management techniques, then you are more likely to be given the opportunity to lease/use that additional land than someone who is not doing so. For example, it is not uncommon for neighbours that have land they are not using to offer it to someone with grazing animals. However, they are unlikely to do this if they see that the land you are currently using is badly managed.

In this situation you may want to use facilities/materials that can be removed and taken with you if you ever move on. There are various options for temporary/relocatable shade/shelters (that have the added advantage of not usually requiring planning permission), fences (including sectional holding yard fences) and even surfaces (such as rubber paving mesh).

—

See the third book in this series *The Equicentral System Series Book 3 – Horse Property Planning and Development* for lots of ideas and solutions.

—

If you need more land and you are already demonstrating good land management techniques, then you are more likely to be given the opportunity to lease/use that additional land.

On a livery yard (boarding/agistment facility)

It is perfectly possible to have this system in operation on a horse livery yard. If the horses already live in herds, then the issues are just the same as for setting this system up on a private-use property. If they do not, then first you have to establish the logistics of how you will integrate the horses into herds. There are several options. You may decide to have a mare group and a gelding group (or several depending on the numbers). You may have a variety of groups, for example, you may decide to let owners group their horses so that they are able to share horse-keeping duties with friends.

Hopefully, the property does not currently have a single horse shade/shelter in every paddock as these will be too small for grouped horses.

Small paddocks that previously housed one horse each can now be rotationally grazed. A central holding yard area will need to be constructed for each herd.

If you wish to implement this system and you would prefer that the owners do not enter a paddock containing a large herd of horses, then you can create a routine whereby the horses come into individual areas once or even twice a day (preferably these areas should lead off from the large surfaced holding yard). This can be very useful in cases of horses receiving different levels of supplementary feed etc.

With single horses in 'private paddocks'

Please note: we are not advocating keeping horses separate to each other, but some owners will *never* put their horse with another horse and very occasionally there are good reasons for separating horses.

Separated horses can still benefit from better pasture management and a better shade/shelter arrangement that allows some socialisation, if it is not already in place.

Horses in 'private paddocks' should have access to a shade/shelter, which should be positioned at the gateway (within a surfaced holding yard) and alongside the next 'private paddock', so that two horses can socialise.

Horses in 'private paddocks' should have access to a shade/shelter, which should be positioned at the gateway (within a surfaced holding yard) and alongside the next 'private paddock', so that two horses can socialise.

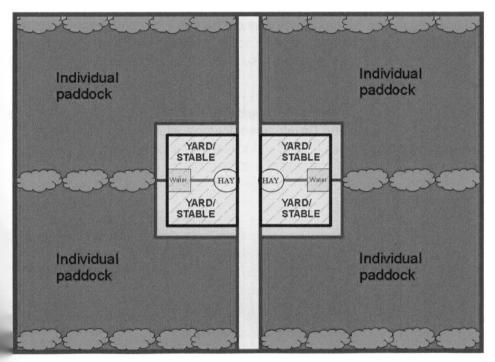

We do not advocate horses socialising over fences, but in this situation, if the partition between the two horses is solid from their chest down (so that they cannot injure a leg) and open above chest height (preferably totally open rather than 'caged'), then two separated horses can and will happily spend many hours 'hanging out' in this area rather than on the sensitive pasture.

When the weather is too wet or too dry the horses can be temporarily prevented from going out of the yard and causing damage that will lead to less quality grazing in the future.

By subdividing any pasture that is available to such horses (as long as this does not create ridiculously small areas) and rotating them around these areas, the land has time to rest and recuperate, resulting in better quality grazing for the horses into the future. For the pasture, any rest is better than none.

See the section *Temporary laneways* for ideas about subdividing smaller/awkward areas of land.

Starting from scratch

If you are in the fortunate position of being able to start a horse facility from scratch, you have much planning to do.

Setting up **The Equicentral System** from scratch should cost less than setting up conventional facilities. Money that would be spent on items such as stables, individual shelters etc. can instead be spent on a surfaced holding yard/s with shelters. Don't forget you can create a surface that can be used for riding/training as well!

—

See the third book in this series *The Equicentral System Series Book 3 – Horse Property Planning and Development* for more information about using a surfaced holding yard as a riding surface, and for information about the total planning of a horse property.

—

Minimising laneways

Depending on the layout of the property, it should be possible to minimise laneway usage, particularly if the internal fences are not yet established.

If you are utilising **The Equicentral System** the horses will be living as a herd, moving *themselves* around the property, only ever having access to one paddock at a time, therefore the paddocks should be arranged so that they lead either directly back to the surfaced holding yard or, if this is not possible (in the case of a long narrow property for example), temporary laneways can be constructed (see the section *Temporary laneways*).

Aim to *minimise* laneways for several reasons:

- Laneways take up space that could otherwise be used as pasture for grazing.

- Laneways concentrate hoof activity to a narrow strip and therefore create land degradation problems such as mud/dust, soil erosion, weeds.

- Laneways are difficult to harrow, mow, weed etc.

- Laneways require more fencing and sometimes require surfacing, therefore extra expense.

Laneways take up space that could otherwise be used as pasture for grazing.

possible, create a layout for your land that reduces or eliminates the need
r laneways. There are several ways that you can do this:

You may be able to have your paddocks 'fan-out' from the surfaced holding yard so that all paddocks lead directly back to this area without the need for laneways.

- You can utilise 'temporary' laneways so that the land used as a laneway is only used as such when necessary and becomes part of the paddock again when not needed as a laneway (see the section **Temporary laneways**).

Property A has a laneway leading to the far paddock but if possible, lay the property out so that there are minimal or no laneways (property B).

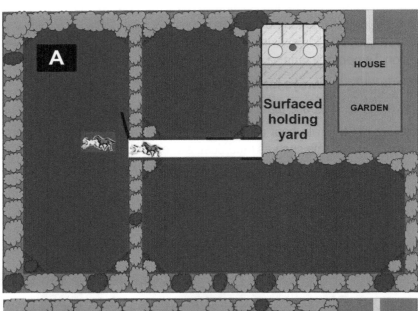

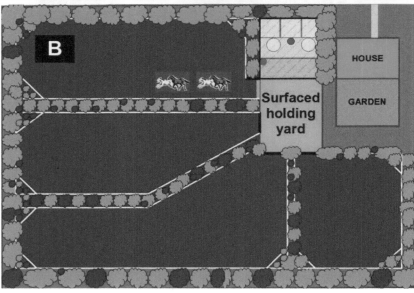

Temporary laneways

A *temporary* laneway can be constructed from temporary electric fence posts (sometimes called 'tread-ins') and electric fence tape to create a 'laneway' that takes the horses to the far end of a long narrow paddock, or even across one paddock to a paddock beyond. This is sometimes preferable to erecting a permanent laneway, because it can be *removed* when the horses are grazing the near section or the near paddock. The land that *would* become a permanent laneway is spared and can be managed as part of the paddock for some of the year; it is far easier to manage land as part of a paddock than to manage land as part of a laneway.

An alternative to using temporary electric fence posts is to put permanent fence posts (e.g. pine poles or steel posts with plastic caps on) in a line where you need them and fasten electric tape carriers to them. This way, an electric tape can be run out through them when necessary and reeled back in when it is not needed.

—

See the third book in this series *The Equicentral System Series Book 3 – Horse Property Planning and Development* for information about fences, including electric fences.

—

A temporary laneway can be used to take horses to a far paddock on a narrow property.

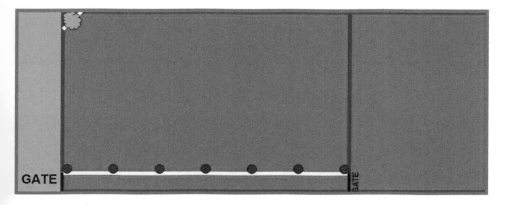

To strip graze a paddock using a temporary laneway, the first stage would look like this....

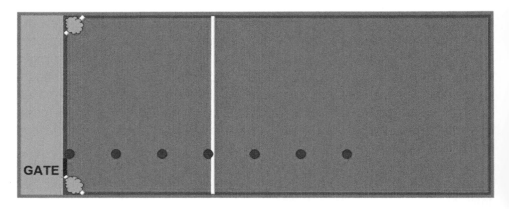

...the second stage would look like this....

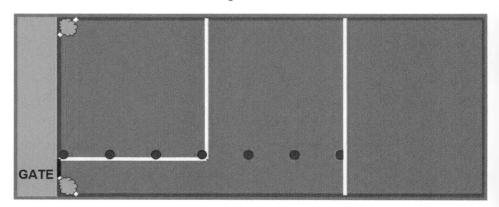

...and the third stage would look like this....

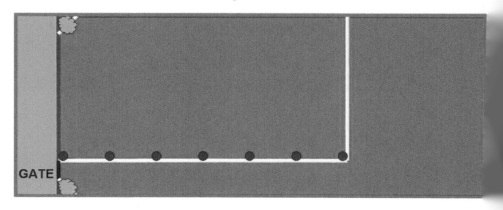

Constructing a holding area

It is imperative that you have a surfaced area for horses to stand. Otherwise, you are quickly going to have mud and dust, soil loss, weeds etc. You are also going to see the skin conditions that are associated with mud such as mud fever/greasy heel etc. If you need to construct a purpose built holding area (rather than utilise something that is already in place) you will need information about this subject.

—

See the third book in this series *The Equicentral System Series Book 3 – Horse Property Planning and Development* for lots more information about constructing a surfaced holding yard.

—

Constructing a shade/shelter

It is imperative that you have shade/shelter for your horse/s. This will increase their need to move themselves back to the holding area and is important for protection both from inclement weather and from insects. There are a huge variety of options ranging from traditional to non-traditional, and from permanent to temporary/relocatable.

—

See the third book in this series *The Equicentral System Series Book 3 – Horse Property Planning and Development* for lots more information about shade/shelters.

—

Fencing considerations

As a general rule, your external (perimeter) fence and any areas in which horses are being confined in a smaller space should have good solid permanent fencing. Anywhere that horses can move away from each other can, if necessary, be fenced inexpensively with electric fencing, certainly initially. Avoid having electric fencing around the holding yard if possible and, if you do use it in laneways, be aware that horses can knock each other into it and it can therefore be stressful for horses when they cannot get out of each other's way.

—

See the third book in this series *The Equicentral System Series Book 3 – Horse Property Planning and Development* for lots more information about fencing.

—

Management solutions

Feeding confined horses

This book does not cover the subject of feeding horses in detail, but this section gives some pointers to keep in mind for confined horses.

The term 'ad-lib' means that something is provided on an 'all you can eat' basis. In the case of hay provision, it means that a horse always has hay available as opposed to being fed measured amounts. It might sound crazy to feed a horse 'ad-lib', but this is what a horse has evolved to deal with. In the naturally-living situation, they are surrounded by their food and graze in bouts and with periods of rest, rather than eating a 'meal' as a predator does and then having to go without food until they make another kill.

In the domestic situation, we can more closely copy this natural situation of having ad-lib feed by aiming to have low energy ad-lib hay or pasture available for our horses.

Horses that are confined, and therefore unable to graze, must be provided with plenty of fibre to make up for not being able to graze. – remember – without fibre acid builds up in the stomach.

One of the most common (and deadliest) mistakes made by horse owners is to feed their horse as they would feed themselves or their dog - on small but high energy meals. Humans and dogs naturally eat much smaller amounts of higher energy food (relatively). This is because their food types are relatively higher in energy (meat and relatively easy to digest vegetables etc.). Horses are completely fferent; their food (pasture plants) is difficult and time consuming to digest and erefore, confined horses should be provided with enough hay to allow them to raze' as and when they want. As already mentioned, ideally, hay should be ovided on an 'ad-lib' (an 'all-you-can-eat') basis when they are not grazing.

Another common horse management mistake is to 'lock horses up' without food an attempt to reduce their feed intake; this practice is commonly done with rses that are getting fat on pasture. Remember - this is not good horse anagement because it leads to gorging when the horse is allowed to eat again.

Another common horse management mistake is to 'lock horses up' without food in an attempt to reduce their feed intake; this practice is commonly done with horses that are getting fat on pasture.

horses have long periods without food, the risks of colic and gastrointestinal cers increase and even laminitis can be brought on by the stress caused by correct feeding (including 'starving').

Clean but low-energy grass hay is better for feeding horses 'ad-lib'; rather than cerne/alfalfa hay, because it is less nutritionally dense. Therefore, more of it can e eaten, thus satisfying the horse's high frequency chewing rates and the guts ed to be constantly processing fibre.

If a horse tends to get fat easily, aim to reduce the *energy* value in the hay in der to maintain the quantity hay; for these animals, aim to source hay with a low gar value. This can be hard to determine, but if you are buying it from a oduce/feed store, you need to ask if they have hay that has had a basic tritional analysis carried out on it (some produce/feed stores will now provide is service). Soaking suspect hay in water (it can then be fed wet) for at least an ur before feeding will help to leach out some of the sugar content.

Be aware that the results of soaking are variable depending on how much sug there was to start with and the temperature of the water (warm/hot water will lea more sugar). In addition, increase your horse's exercise – this is a very importa but often ignored point. Remember - horses are meant to move a lot. It is comm for people to go to great lengths to reduce the chance that their horse will devel or have a reoccurrence of laminitis – by 'micro managing' the horse diet. It is pa ticularly surprising that many horse owners will opt to buy expensive supplemen and feeds in preference to planning a more naturally active lifestyle for th animals. Increased exercise is a cheaper, more effective way to prevent obes and it's diseases, it also leads to a mentally more balanced horse. (see the secti *Ideas for extra exercise*).

Many people underestimate how much fibre a horse actually needs. An avera hay bale (small square) has 10 biscuits (sections) of hay. If a horse is confined all or most of each day, a medium size (14-15hh) horse needs *at least* 1/3 (3 biscuits) of a (heavy compacted) bale to go through its gut daily. A larger hor needs as much as 1/2 of a bale (5 biscuits) or even 3/4 of a bale (7-8 biscuits) hay per day. This is just a very rough guide, as bales of hay vary very much in weig

Another rough calculation is that a mature horse needs to eat approximately 2 of its bodyweight in Dry Matter (DM) per day. So a 500kg (1100lb) horse will ne 10kg (22lb) of hay (hay does not have much water content so if you are feedi haylage or silage, which does contain water, this figure would be higher). addition, a horse may need minerals adding to their diet.

These amounts are just to give an inexperienced horse owner a rough ide of the volume a horse actually needs. In reality, a horse should always hav access to ad-lib hay when not grazing.

A horse that is working hard may also need supplementary hard feed (e.g. grain or mixes), but be careful as another common mistake that horse owners make i that they tend to overestimate their horse's hard feed requirements.

Remember - horses that are 'group housed' should be able to get out of each other's way and should be separated for supplementary feeding if communal feeding initiates aggression. Horses should ideally be separate into individual yards or stables for the short time that it takes to eat any concentrate feed; both for their own safety and the safety of their handlers

Changing a horse to 'ad-lib' feeding

If your horse has always had measured/restricted amounts of food rather than ad-lib food then you will have to be careful about changing them over. When horses have been withheld from food they tend to 'gorge' when first allowed to eat at will. Remember - a horse would naturally spend most of its day eating fibre, its whole physiology has evolved to allow it to do this efficiently. When you use restrictive feeding/grazing practices, this is in complete contrast to what the horse has adapted to do and when combined with the horses natural instinct to try to gain weight whenever possible, it is easy to see why many horses develop 'eating disorders'.

So, if the horse is currently living on short stressed grasses (and is overweight), it would *not* be a good idea to switch to turning him or her out on long grasses straightaway; even though these grasses are lower in sugar per mouthful, because the horse in question will initially gorge themselves. A better strategy would be either of the following:

Option 1: Over winter – with no access to pasture initially:

- You will need the use of surfaced holding yards – preferably as part of an **Equicentral System**.

- During winter, when the horse is receiving no pasture, feed ad-lib low-energy hay in a surfaced yard – preferably with other horses. You may want to soak this hay in warm water for at least one hour before feeding as a further precaution, particularly if you are not sure what the energy level of that hay is. **An extremely important factor is that the hay does not run out – at all, ever!** This is because, if it does, the horse thinks he or she is being 'starved' again and behaves accordingly (e.g. starts to gorge when food is available again).

- Aim to reduce the horse's weight gradually but significantly over the winter by totally avoiding high energy supplementary feeds, avoiding rugging unless absolutely necessary (but ensure that the horse can get under a shelter) and *increasing exercise* (see the section *Ideas for extra exercise*). You need to aim for a condition score of no more than 2.5 by the start of spring.

Condition scoring

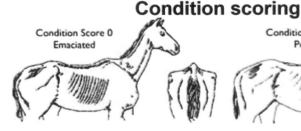

Score 0 (Very Poor). *Neck* - marked 'ewe neck' - narrow and slack at base. *Back & ribs* - skin tight over ribs, very prominent backbone. *Pelvis & rump* - very sunken rump, deep cavity under tail, angular pelvis, skin tight.

Score 1 (Poor). *Neck* - 'ewe' shaped, narrow & slack at base. *Back & ribs* - ribs easily visible, skin sunken either side of backbone. *Pelvis & rump* - sunken rump but skin slacker, pelvis and croup highly defined.

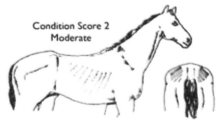

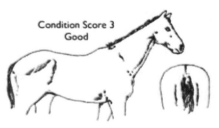

Score 2 (Moderate). *Neck* - narrow but firm. *Back & ribs* - ribs just visible, backbone well covered but can be felt. *Pelvis & rump* - flat rump either side of backbone, croup well defined, some fat, slight cavity under tail.

Score 3 (Good). *Neck* - firm, no crest (except in a stallion). *Back & ribs* - ribs just covered but easily felt, no gutter along back, backbone covered but can be felt. *Pelvis & rump* - covered by fat and rounded, no gutter, pelvis easily felt.

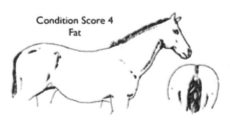

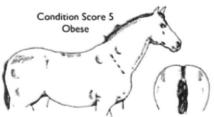

Score 4 (Fat). *Neck* - slight crest. *Back & ribs* - ribs well covered, need firm pressure to feel, gutter along backbone. *Pelvis & rump* - gutter to root of tail, pelvis covered by soft fat - felt only with firm pressure.

Score 5 (Very Fat). *Neck* - marked crest, very wide and firm, lumpy fat. *Back & ribs* - deep gutter along back, back broad and flat, ribs buried cannot be felt. *Pelvis & rump* - deep gutter to root of tail, skin is distended, pelvis buried under fat.

- The horse can still be given minerals etc. if you feel that they are needed, but these do not have to be added to high calorie feed; they can be added to a small amount of chaff.

—

We believe that *not* allowing this naturally occurring weight loss to happen in winter is one of the primary reasons for the obesity epidemic today.

—

- If your land recovers enough to grow pasture before the winter, and this paddock is locked up, you can start to introduce the horse gradually to the pasture over the latter part of the winter. Letting pasture grow long and then allowing horses to harvest it themselves in winter is called 'foggage' or 'standing hay'. This practice has many, many advantages including that it saves the costs of cutting, baling and storing hay (and the risk of it 'failing' as a crop). By mid-winter, it will have more fibre value than nutritional value; in other words, it is ideal for a 'weight challenged' horse.

—

See the second book in this series *The Equicentral System Series Book 2 - Healthy Land, Healthy Pasture, Healthy Horses* for more information about this practice.

—

- When the rest of your land is ready to receive horses again in the spring, you can gradually allow the horse in question to have at first one grazing turnout session (grazing bout) per day over a period of about a week (in addition to ad-lib hay), then allow this session to be longer (for about a week) and so on.
- Extra exercise may be necessary during this period too and it is extremely beneficial if you can do this (see the section *Ideas for extra exercise*).
- By using **The Equicentral System** you will be able to, at first, dictate when that first grazing period takes place; very late evenings (well after sundown) or very early mornings are a good time – but you will need to bring them back in by lunch time at the latest. Initially, avoid allowing the horse to graze between mid-day and nightfall, because this is when the sugars in the grasses are at their highest levels.
- By late spring/early summer, as long as you are keeping a good watch on the horse's body condition score and are not reverting to restricting their low energy hay intake, you should be able to allow night and day grazing bouts with free access to the paddock that is currently in use.

- By the time the pasture is growing higher energy feed in the spring, the hors will have relaxed and will not be as tempted to gorge. If you have carried out th above steps, the horse will now have a much lower body condition score an will be in a much safer position. You will need to keep up this pattern of reducin the horse's weight every winter and keeping up the extra movement whenever is necessary.

An extremely important factor is that the hay does not run out – at all, ever!

Option 2 – During summer – with access to pasture.

- This option is if you would like to start changing a horse over to ad-lib feedin right away, without waiting for winter. Again you will need the use of surface holding yards – preferably as part of an **Equicentral System**.

- Initially, confine the horse by day, on ad-lib low energy hay. Again, it i imperative that the hay **never** runs out and again, you may want to soak th hay in water before feeding, only allowing one grazing turnout period per day a per the previous example.

- You will need to closely monitor the horse's weight and you should **definitel** increase exercise during this period, which should be easier for you at this tim of year (see the section *Ideas for extra exercise*).

As in the above option, avoid supplementary feeding and rugging. Make sure the horse has access to shade/shelter and they should preferably be kept with other horses. Carry on adding grazing time as per option 1, but **only** if you feel the horse is not increasing weight too fast.

The idea is that you are initially controlling the horse's intake by allowing ad-lib access to low energy hay, but you are switching the horse over to not feeling restricted *at all*. Remember - restricted feeding can actually increase insulin resistance levels because the body reacts by going into 'starvation mode' - **never lock a horse up without something to eat.**

Never limit hay, limit grazing time initially if you feel the horse is gaining weight too fast. When winter arrives and for every winter from now on, still aim for the horse to lose some weight, because this is what the horse has done for naturally for eons; lost weight in winter and not been in as dangerous a position when the feed quality increases in the spring. They can then relatively safely gain some weight gradually during spring and summer. This is a better strategy than trying to maintain the horse's weight at exactly the same weight all year round.

You may need to learn more about pasture plants too, including factors that make them safer, or not as safe, to graze.

Make sure the horse has access to shade/shelter and they should preferably be kept with other horses.

Source - Gina Sykes of Bathurst Equine Agistment - Aus.

Before embarking on any radical changes such as those outlined abov have the horse checked out by an experienced *equine* veterinaria preferably someone who has a particular interest in the subject of equi obesity. You could also engage an equine nutritionist; preferably a *independent* equine nutritionist.

—

See the second book in this series *The Equicentral System Series Book 2 - Healthy Land, Healthy Pasture, Healthy Horses* for more information.

—

Ideas for extra exercise

Curiously, most pet owners recognise that if they own a dog they should 'walk' (even if they do not actually do it) but many horse owners do not apply the san ethos to horsekeeping. Maybe because dogs live in or around a home and tend remind their owners that they want to go out? Horse owners tend to assume th their horse/s get enough exercise if they are tuned out on a pasture but as we nc know they usually need more exercise than that, plus they are usually receivi more energy (from pasture/feed) than they are using.

There are various ways, apart from general riding, that you can incorpora *extra* exercise into a horse's management routine.

- Driving - small ponies in particular can be trained to drive. Driving is a go pastime for horses and people.

- Hand walking – there is no reason why you cannot take a horse for a wa smaller ponies in particular are not difficult to do this with. You could even wa the dog at the same time!

- Running – some owners jog/run with their horse in hand for mutual fitness benefi

- Hill climbing in hand – if you live in a hilly area you can walk/jog/run the hills a you can use your horse to help you up the hills by holding onto their mane.

- Lunging – just ten minutes a day of trotting on the lunge is great exercise.

- Round penning – like lunging but loose in a round yard.

- 'Riding and leading' – if you are riding anyway, why not lead another horse wh doing so?

You (and your horse) may need instruction before carrying out some of the activities, but the benefits will be worth it.

Introducing horses to herd living

Careful introduction of a horse into a herd will vastly reduce any risks and an added bonus is that grazing management is much easier when horses live together, because they can be rotated around areas as a group. One horse per paddock does not allow the pasture any rest and recuperation time; a welfare issue for grass! In fact, 'managing' pasture in this way leads to stressed grass that is not good for horses and land degradation problems.

—

We cover this subject in detail in the second book in this series *The Equicentral System Series Book 2 - Healthy Land, Healthy Pasture, Healthy Horses*.

—

If you decide to integrate your horses into a herd because of the horse welfare *and* land management benefits, there are several things to think about and steps to take so that the integration goes smoothly. First of all, think about each of the horses in question and decide if it will be best to have one herd or more than one herd. To end up with just one herd is the best outcome because this will be much easier for land management, but this may not be possible in your situation.

Some of the factors that will help you to decide include the age and sex of the individual animals; for example, young and boisterous horses may be too energetic for a *very* old horse whereas, older horses, on the other hand are usually very good at holding their own with younger horses up until a certain age (which is different for all horses), when they may start to need more specialised care in general.

It generally works best if there are more mares than geldings in a herd, because some geldings still have some entire (stallion) behaviour and can become protective of mares to the point that they will chase other geldings if mares are present. So, if you decide to have two herds for example, it may be better in this case to have one gelding with the mares in one herd and the rest of the geldings in another herd. This is similar to the natural groups that occur in the naturally-living situation e.g. a stallion with some mares and a bachelor group consisting of males of all ages. All horses are different though and various scenarios can work.

There are many ways that a new horse can be introduced to a group of horses. Remember that the existing group will have a social structure and the introduction of a new member will temporarily disrupt this. It is simply not safe to turn the new member out into the group and 'let them get on with it'. In a confined space, the new horse can be run into or over a fence by the other horses. It is better to let the new horse get to know at least one member of the group in separate, securely fenced yards (preferably 'post and rail') or stables that have an area where two horses can safely interact with each other. This way, the newcomer can approach the fence or wall to greet the other horse, but can also get away if necessary. There will usually be squealing, but this is perfectly normal behaviour when horses meet and greet each other.

It is better to let the new horse get to know at least one member of the group in separate, securely fenced yards (preferably 'post and rail') or stables that have an area where two horses can safely interact with each other.

Once these two horses are accustomed to each other, you can turn them out together and then add other herd members gradually one at a time. Try not to give them any hard feed (if the horses are being supplementary fed), only hay, just before you turn them out so that they get down to grazing sooner. It is safer if the horses are left unshod at least for the initial introductions. Hoof boots can also be used. Make sure that resources are plentiful and easy to reach, for example situate the water away from a corner so that they can each drink safely and no one gets trapped. The group must be watched very carefully during these times.

Keep in mind that you will usually see the most excitable behaviour between the horses in the first hour or so after turning them out together. It can be challenging when introducing horses to one another for the first time, but as long as it is done in a safe manner, usually after a short, somewhat noisy and lively period things settle down. There is bound to be some initial excitement when two horses meet for the first time, particularly if they have been previously kept alone or are new additions to an established herd. Think about when introducing new cat or dog t an established household, there are normally one of two skirmishes until everyone gets used to each other. This is no different in the horse world, but because they are large, energetic, noisy and valuable animals we are often shocked by the first encounter. We have to be aware that some of that behaviour is horseplay; by its definition in human terms, loud and boisterous, but remember this is what horses do. However, we have to be aware of the difference between play behaviour and aggressive behaviour, and ensure that we monitor the situation. With some horses this can be very difficult, as they have been mentally damaged by the way they have been raised or separated in the past.

The Equicentral System - in conclusion

As we have shown, there are many challenges facing modern horse owners. Traditional systems are not meeting the needs of our modern horses. We have to look at more holistic management systems that can work within the boundaries and limitations of our and that of horse's lifestyles.

The simplest way to achieve this is to accept that although we cannot provide a fully natural lifestyle for our domesticated horses, we can learn from nature and work with it rather than against it. Once we start to do this, everything becomes easier, more productive and healthier. We should also invest time in learning about what our horses actually need, this way we are better equipped to make informed choices not only about how we manage our horses but also in who we turn to for advice when needed.

By becoming more responsible, sustainable and ethical horse owners we can ensure that we strive to create an environment which is conducive to creating healthy land, which then creates healthy pasture to ensure we have healthy horses.

Further reading - A list of our books

Buying a Horse Property

Buying a horse property is probably the most expensive and important purchase you will ever make. Therefore, it is very important that you get it right. There are many factors to consider and there may be compromises that have to be made. This guide to buying a horse property will help you to make many of those very important decisions.

Decisions include factors such as whether to buy developed or undeveloped land? Whether to buy a smaller property nearer the city or a larger property in a rural area? Other factors that you need to think about include the size and layout of the property, the pastures and soil, access to riding areas, the water supply, and any possible future proposals for the area. These subjects and many more are covered in this book.

A useful checklist is also provided so that you can ask the right questions before making this very important decision.

If you are buying a horse property, you cannot afford to miss out on the invaluable information in this book!

The Equicentral System Series Book 1: Horse Ownership Responsible Sustainable Ethical

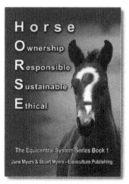

With horse ownership comes great responsibility; we have a responsibility to manage our horses to the best of our ability and to do this sustainably and ethically.

Horse keeping has changed dramatically in the last 30 to 40 years and there are many new challenges facing contemporary horse owners. The modern domestic horse is now much more likely to be kept for leisure purposes than for work and this can have huge implications on the health and well-being of our horses and create heavy demands on our time and resources.

We need to rethink how we keep horses today rather than carry on doing things traditionally simply because that is 'how it has always been done'. We need to loo at how we can develop practices that ensure that their needs are met, without compromising their welfare, the environment and our own lifestyle.

This book brings together much of the current research and thinking on responsible, sustainable, ethical horsekeeping so that you can make informed choices when it comes to your own horse management practices. It starts by looking at the way we traditionally keep horses and how this has come about. It then discusses some contemporary issues and offers some solutions in particular a system of horsekeeping that we have developed and call **The Equicentral System.**

For many years now we have been teaching this management system to horse owners in various climates around the world, to great effect. This system has many advantages for the 'lifestyle' of your horse/s, your own lifestyle and for the wider environment - all at the same time, a true win-win situation all round.

The Equicentral System Series Book 2: Healthy Land, Healthy Pasture, Healthy Horses

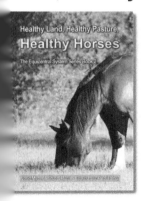

If you watch horses grazing pasture, you would think that they were made for each other. You would in fact be correct; millions of years of evolution have created a symbiotic relationship between equines (and other grazing animals) and grasslands. Our aim as horse owners and as custodians of the land should be to replicate that relationship on our land as closely as possible.

In an ideal world, most horse owners would like to have healthy nutritious pastures on which to graze their horses all year round. Unfortunately, the reality for many horse owners is far from ideal. However, armed with a little knowledge it is usually possible to make a few simple changes in your management system to create an environment which produces healthy, horse friendly pasture, which in turn leads to healthy 'happy' horses.

Correct management of manure, water and vegetation on a horse property is also essential to the well-being of your family, your animals, your property and the wider environment.

This book will help to convince you that good land management is worthwhile on many levels and yields many rewards. You will learn how to manage your land in a way that will save you time and money, keep your horses healthy and content *and* be good for the environment all at the same time. It is one of those rare win-win situations.

The Equicentral System Series Book 3: Horse Property Planning and Development

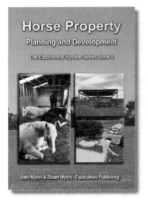

It does not matter if you are buying an established horse property, starting with a blank canvas or modifying a property you already own; a little forward planning can ensure that your dream becomes your property. Design plays a very important role in all our lives. Good design leads to better living and working spaces and it is therefore very important that we look at our property as a whole with a view to creating a design that will work for our chosen lifestyle, our chosen horse pursuit, keep our horses healthy and happy, enhance the environment and to be pleasing to the eye, all at the same time.

Building horse facilities is an expensive operation. Therefore, planning what you are going to have built, or build yourself is an important first step. Time spent in the planning stage will help to save time and money later on.

The correct positioning of fences, laneways, buildings, yards and other horse facilities is essential for the successful operation and management of a horse property and can have great benefits for the environment. If it is well planned, the property will be a safer, more productive, more enjoyable place to work and spend time with horses. At the same time, it will be labour saving and cost effective due to improved efficiency, as well as more aesthetically pleasing, therefore it will be a more valuable piece of real estate. If the property is also a commercial enterprise then a well-planned property will be a boon to your business. This book will help you make decisions about what you need, and where you need it; it could save you thousands.

Horse Properties - A Management Guide

This book is an overview of how you can successfully manage a horse property sustainably and efficiently. It also complements our one day workshop - *Healthy Land, Healthy Pasture, Healthy Horses*.

This book offers many practical solutions for common problems that occur when managing a horse property. It also includes the management system that we have designed, called - **The Equicentral System**.

This book is a great introduction to the subject of land management for horse keepers. It is packed with pictures and explanations that help you to learn, and will make you want to learn even more.

Some of the subjects included in this book are:

The grazing behaviour of horses.
The paddock behaviour of horses.
The dunging behaviour of horses.
Integrating horses into a herd.
Land degradation problems.
The many benefits of pasture plants.
Horses and biodiversity.
Grasses for horses.
Simple solutions for bare soil.
Grazing and pasture management.
Grazing systems.
Condition scoring.
Manure management... and much more!

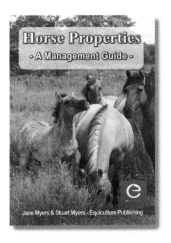

Horse Rider's Mechanic Workbook 1: Your Position

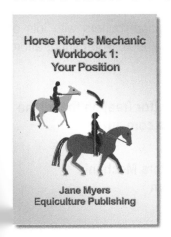

Many common horse riding problems, including pain and discomfort when riding, can be attributed to poor rider position. Often riders are not even aware of what is happening to various parts of their body when they are riding. Improving your position is the key to improving your riding. It is of key importance because without addressing the fundamental issues, you cannot obtain an 'independent seat'.

This book looks at each part of your body in great detail, starting with your feet and working upwards through your ankles, knees and hips. It then looks at your torso, arms, hands and head. Each chapter details what each of these parts of your body should be doing and what you can do to fix any problems you have with them. It is a step by step guide which allows you to fix your own position problems.

After reading this book, you will have a greater understanding of what is happening to the various parts of your body when you ride and why. You will then be able to continue to improve your position, your seat and your riding in general. This book also provides instructors, riding coaches and trainers with lots of valuable rider position tips for teaching clients. You cannot afford to miss out on this great opportunity to learn!

Horse Rider's Mechanic Workbook 2: Your Balance

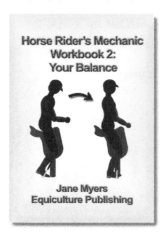

Horse Rider's Mechanic
Workbook 2:
Your Balance

Jane Myers
Equiculture Publishing

Without good balance, you cannot ride to the best of your ability. After improving your position (the subject of the first book in this series), improving your balance will lead to you becoming a more secure and therefore confident rider. Improving your balance is the key to *further* improving your riding. Most riders need help with this area of their riding life, yet it is not a commonly taught subject.

This book contains several lessons for each of the three paces, walk, trot and canter. It builds on **Horse Rider's Mechanic Workbook 1: Your Position**, teaching you how to implement your now improved position and become a safer and more secure rider.

The lessons allow you to improve at your own pace, in your own time. They will compliment any instruction you are currently receiving because they concentrate on issues that are generally not covered by most instructors.

This book also provides instructors, riding coaches and trainers with lots of valuable tips for teaching clients how to improve their balance. You cannot afford to miss out on this great opportunity to learn!

You can read the beginning of each of these books (for free) on the on the Equiculture website www.equiculture.com.au

We also have a website just for Horse Riders Mechanic www.horseridersmechanic.com

Most of our books are available in various formats including paperback, as a PDF download and as a Kindle ebook. You can find out more on our websites where we offer fantastic package deals for our books!

Make sure you sign up for our mailing list while you are on our websites so that you find out when they are published. You will also be able to find out about our workshops and clinics while on the websites.

Recommended websites and books

Our websites **www.equiculture.com.au** and **www.horseridersmechanic.com** have extensive information about horsekeeping, horse care and welfare, riding and training. Please visit them and you will find links to other informative websites and books.

Bibliography of scientific papers

Please go to our website **www.equiculture.com.au** for a list of scientific publications that were used for this book and our other books.

Final thoughts

Thank you for reading this book. We sincerely hope that you have enjoyed it. Please consider leaving a review of this book at the place you bought it from, or contacting us with feedback, stuart@equiculture.com.au so that others may benefit from your reading experience.